To Succeed Where Others Failed

The Untold Story of the Marshall Plantation Raid

Bruce Seaman

Bruce Seaman, Ocala, Florida

To Succeed Where Others Failed: The Untold Story of the Marshall Plantation Raid by Bruce Seaman

Copyright © 2022 Bruce Seaman

All rights reserved. No portion of this book may be reproduced in any form without permission from the publisher, except as permitted by U.S. copyright law. For permissions contact: bruceseaman@zoho.com

Library of Congress Control Number: 2022918067

ISBN: 979-8-9870018-0-6

Cover design by Rachel Seaman

Cover photo is a unit of the 4th US Colored Troops of Washington, DC in November 1865 - public domain.

Printed in the USA

Third Edition

(updated August, 2024)

Contents

Acknowledgements	1
Author's Preface	3
A Note on the Third Edition	11
A Note About The Maps	13
Introduction	15
1. Black Men Not Wanted	24
2. The Black Soldier "Experiment" Begins	32
3. Black Soldiers in Action in the South	43
4. Black Soldiers in Action in Florida	51
5. Confederate Captain J. J. Dickison	66
6. Why Raid the Marshall Plantation?	81
7. Critical Intelligence and a Unique Strategy	87
8. Getting Started	98
9. The First Leg of the Journey	117
10. The Raid	122
11. The Chase Begins	140

12.	The Last Leg of the Journey	160
13.	Commentaries After the Raid	171
14.	Conclusion	181
15.	Postscript	190
Brief Bibliography		219
Catalog of Images		222
Appendices		226

Acknowledgements

This book would never have been written without the journalistic endeavors of Rick Allen. His article published in *Ocala Star-Banner* in February 2016 went above and beyond routine reporting.

In chasing down clues and leads in pursuit of a fuller version of the Marshall Plantation raid, he discovered the letter of Sergeant Henry Harmon who gave a full account of the raid in an April 1865 edition of *The Christian Recorder*, a popular African Methodist Episcopal (AME) church publication. The letter goes into considerable detail in describing the raid, surely the testimony of a raid participant, and providing a primary source that could be compared to the accepted version of the raid provided by Confederate officer and author J.J. Dickison and elaborated by Confederacy apologists. The comparison between these primary sources in this book aims to reveal a less biased and more accurate account.

Rick Allen

A tip-of-the-hat is also deserved for Rick's editor at the time, Jim Ross, who okayed the publication of Rick's exceptionally long 2400-word article. If Rick had done the work and the article had never been published, I would never have read it or gotten to writing this book.

I will also acknowledge my wife and my family for their forbearance as I have droned on about different aspects and discoveries along the way, and for their critical commentary and helpful guidance that aided me in bringing this to fruition.

Many others have also encouraged me, like local historian Cynthia Graham and her longtime sidekick Monica Bryant, and the folks at *Master the Possibilities*, the adult learning center at the retirement community *On Top of the World* where I have made regular presentations about the raid, allowing me to hone and develop the story.

A final acknowledgment is due to Ed McLaughlin whose work in compiling an online database of documentation from the National Archives and other research for all of the Black soldiers who were trained at Camp William Penn in Philadelphia, including the 3rd US Colored Troops who participated in the raid, is truly a labor of love and a remarkable commitment. The Google Drive database is named CROHL, abbreviated "Citizens for the Restoration of Historical La Mott," dedicated to the tens of thousands of Union soldiers who trained at Camp William Penn in Philadelphia. That link is:

https://drive.google.com/drive/folders/1gjvLtr7LSdkvBQ_kStwdwvtr-rO9K9qb

Author's Preface

This story of the Marshall Plantation raid became a matter of fascination to me after reading journalist Rick Allen's discoveries in his *Ocala Star-Banner* article in February 2016 that changed the whole nature of the story.

By surfacing a Union account, testimony from a first-hand participant recorded by the letter writer, written only weeks after the raid, Rick revealed a whole other side to the story that had not been presented by local Confederacy apologist historians. If there were questions raised by the accepted biased version of the locals, then new questions were raised by having a detailed account from the other side, the Union side.

Questions? Oh my. Let us start with the fact that this raid makes no sense from any apparent perspective. It is so off-the-charts nutty that one would have to dismiss the raiders as collectively insane or collectively stupid. I was unwilling to accept either of those options. Finding a sane and intelligent perspective to make sense of this operation became my obsession and this book.

I am not a scholar-historian and I am not even a Civil War buff *per se*, although I minored in history in college. I do enjoy history and sharing a good story. Storytellers are historians, so that makes all of us historians in some sense.

This is a *really* good story, that is when both sides get represented. Once this exciting story is set in its proper context in the experience of the Black soldier, it can truly be understood as a milestone event in the difficult journey for Black soldiers to be recognized as complete soldiers, leaders, and strategists.

One of my concerns is that often only one side gets represented in portrayals of the Civil War South in contemporary Southern depictions. This is one of those stories that has been intentionally misrepresented. This kind of "white-washing" of stories relevant to Black history and the history of other minorities has been common in accounts of US history. Sadly, it continues today. Since local historians specifically identified the Union's actors in this raid, they had access to information that would portray the events very differently. Yet choices were made to present a flawed and corrupt narrative of "Lost Cause" history.

"Lost Cause" history refers to a revisionist slant with beginnings in the post-Reconstruction period of the late 19th century which sought to redeem the Confederacy, making it appear as an honorable, dignified, thoughtful, and not-so-racist traditional society to cover up the Confederacy's truly despicable system of racial enslavement, White supremacy, hatred, violence, exploitation, and thorough brutality. Echoes of this ridiculous posturing can be heard in silly statements that claim slavery was not the cause of secession and the ensuing Civil War. This claim is made despite the clear record over the question of new states being admitted as either slave or free states, and that preserving slavery and White supremacy were the reasons for secession spelled out in many secession proclamations. Despite its bankruptcy, "Lost Cause" angles have been wildly successful in corrupting the hearts and minds of USAmericans to this day.

I am a retired Presbyterian minister living in Marion County, Florida, the site of the historic raid, who is also interested in the critical, scholarly reading of biblical literature. Here, too, I am not a biblical scholar *per se*, but I do employ critical methodologies in seeking to understand biblical texts. These methodologies expect the raising of difficult questions and wrestling with them to discern the best possible understanding of a text. There should be no "sacred cows" to disrupt the exercise of critical thinking, only a commitment to getting the best possible understanding of what is presented.

I will use this type of critical thinking methodology in reviewing and evaluating the primary texts (and others) about the Marshall Plantation raid, seeking to derive the best possible understanding of what happened and why.

We have two primary texts and several supplementary texts.

For primary texts, first, there is the written report of Confederate Captain J. J. Dickison, Company H, 2nd Florida Cavalry to his commanding officer dated March 20, 1865. It is largely reproduced in the biography published in 1890 by his wife, Mary Elizabeth Dickison, *Dickison and His Men*, and in his own writing of the 11th volume of *Confederate Military History* concerning the war's actions in Florida, published in 1899 (See Appendix for the full text).

Second, there is the letter written by Sgt. Henry S. Harmon. The Sergeant was part of Company B, 3rd US Colored Troops (USCT). Given the lack of first-person references, it would appear that Sgt. Harmon dutifully recorded the detailed account from a first-hand participant, one of the raiders from Company B, 3rd USCT, like him. It seems very doubtful that Sergeant Harmon was a raider himself. (See Appendix for the full text).

He wrote his letter on April 3, 1865 detailing the raid, and the letter was published in the April 22, 1865 edition of *The Christian Recorder*, the official publication of the African Methodist Episcopal Church (AME) founded and headquartered in Philadelphia where the 3rd USCT mustered. During the Civil War, this AME publication reached a wide audience among Black troops throughout the country.

Both primary text accounts have their biases, inconsistencies, exaggerations, and flaws.

First, neither writer was concerned with an accurate chronology, causing the analyst of the raid to calculate and conjecture timelines as well as sort out details that are not presented by the sources in a helpful linear fashion. Indeed, it seems that, in Dickison's case, he conflated different information with different events, and may have sought to dramatize actions which we will confront below.

Second, each was writing for different audiences, and those biases need to be accounted for.

Sergeant Harmon was addressing a Black audience that included a large number of Black Union soldiers nationwide who were sympathetic to the Union cause, depicting an exceptional account of heroism, strategy, leadership, and determination by Black Union soldiers.

Captain Dickison's account is a report to his superior officer, Lieutenant-Colonel W. K. Beard in Tallahassee. Dickison is portraying his account of difficult events with a negative outcome in the best possible light.

Third, the numbers provided are not always reliable with each side overestimating the numbers of the other. This pertains to force levels, casualties, and the like. Such misconceptions were commonplace in Civil War accounts on both sides. (It still happens routinely in modern times.) As a rule of thumb, whatever one side reports as its own numbers is more likely to be accurate than the other side's portrayal.

Fourth, later reports of the second-hand variety transformed the raid on their way to entering into the record.

For example, an account of the raid from the Confederate-oriented perspective by someone identified as "E.O." – quite possibly one of the large-plantation-owning members of the Owens clan in Marion County – was given in the *Quincy (FL) Dispatch* newspaper on March 22, 1865. It gets quickly carried away with itself, with gems like describing the raiders as "cavalry" – not so! Dickison is described inaccurately as capturing "several deserters" – again, inaccurately – in the same severely flawed narrative.

From the Union-oriented perspective, a letter appeared in *The Christian Recorder* on April 29, 1865, a week after Sgt. Harmon's letter was published, which is replete with confused events, factual errors, and exaggerations.

A pro-Union report published in the *Florida Union* of Jacksonville on March 19 is so riddled with misinformation that it is difficult to believe it describes the same raid. It may have been repeated with a second (juicy) report of a raid on a sugar plantation at that time, which is found nowhere else, sowing complete confusion.

These are prime examples of how quickly an account can morph in the second-hand (or third- or fourth-hand) retelling that transforms the events. (See Appendix for the full texts)

Fifth, once such severely flawed narratives enter the record, these flaws may get uncritically replicated in subsequent accounts and narratives into modern times, particularly if they serve the "Lost Cause" apologist narrative.

Quite simply, in analyzing the texts, trying to keep the context in focus, and sorting through the possibilities to define probabilities, the reader will find me posing numerous questions, suggesting possible answers, and naming preferred answers to those questions. Such possible and preferred answers should never be taken as definitive and factual. Rather they are presented as the best understandings after weighing the possibilities. Opinions may vary.

The book is structured in four sections.

The first section, chapters 1-5, includes an introduction to the issues surrounding this raid before proceeding into a general summary of events concerning the Black soldier in the Union army as this 'experiment' in utilizing Black soldiers unfolds.

The focus then shifts to the two units involved with the raid, the 3rd US Colored Troops (3rd USCT) and the 34th US Colored Troops (34th USCT, formerly the 2nd South Carolina Volunteer Infantry or 2nd SCVI). We broadly follow their movements from their beginnings, their deployment in the Union's Department of the South to their deployment in Florida.

This preparatory exploration closes with a summary describing the notorious Confederate cavalry Captain J. J. Dickison, the nemesis of Union operations in northeast Florida.

The second section, chapters 6 and 7, forms the context for the raid, asking questions that beg to be answered about its origins, purpose, and strategy. It closes by outlining the likely process in coming to the raider unit's composition and the approval for the unusual raid to proceed.

The third section, chapters 8-12, reviews the raid chronologically and geographically in considerable detail from its beginning to its ending, based principally on the primary source materials.

The fourth section, chapters 13-15, reviews commentaries about the raid and what happened afterward as well as a postscript that looks at what happened with the key players after the war ended and how the raid has been accounted for historically.

Finally, **let me invite you to share a reader review when you are done.** Your shared opinion will help others decide if this is the kind of book that they want. Indeed, a reader review may have helped you to decide. Thank you in advance!

I hope you enjoy this account as much as I have enjoyed presenting it to groups and writing about it here.

Bruce Seaman

Ocala, Florida

A Note on the Third Edition

If you have already read either the first or second editions, you can find all of the updated information contained in this third edition on my website for free. You don't need to buy the book again! Go to **https://www.bruceseaman.com/updates** and see what has been added or changed.

On the website, there are pieces on the Note to the Second Edition, the White soldier, the Holly Plantation operation, more info about guerilla groups in Florida, and further information about Sgt. Harmon, the letter writer, including an addition to the Appendix with the text of a letter Harmon wrote, published in *The Christian Recorder*, about the harsh punishments for Black soldiers written just before the Jacksonville Mutiny described in the Postscript.

While the book may have been published over a year ago, my interest in the subject and aspects surrounding the raid and its participants has not waned. As new material, resources, or insights are gleaned, they are posted on my website for all to see freely.

This third edition adds the material, resources, and insights posted on the website into the book. The ability to add footnotes/endnotes in my program has allowed me to include additional material as well as transfer some of the laborious details from the text into footnotes/endnotes.

Hopefully, the changes are improvements and make for a better reading about this amazing event and the context in which it occurred.

January, 2024

A Note About The Maps

Maps of Florida in the 19th century were of poor quality; it is a large state and was thinly settled at that time. Poor map quality would be a problem that hampered Union operations regularly. It has also been an issue for providing the reader with something worthy in maps to follow the progress of raid events in the third section.

The geographical dimensions of the narrative embraced by this book require a map since even Floridians have difficulty with the distances involved in this large state. Given the poor quality of Civil War-era Florida maps, particularly its sparsely-populated interior, choices needed to be made.

There is a map of Florida dated 1865 with reasonably good detail for a statewide map and reasonably good accuracy. However, magnification makes it highly problematic to read, a key element for a reader. Labels were added to the selections from this map to highlight the key locations mentioned at various places. The reader should be able to follow the movements as they unfold.

The last map focuses on the last legs of the raiders' flight to St. Augustine and their pursuers' route to intercept them. This map is also from 1865 and focuses on St. Johns County where St. Augustine is located.

A host of other maps were consulted to identify likely routes taken by the raiders and their pursuers.

Of course, this is guesswork. The first-hand sources give little indication of land routes taken.

Above all, the maps are provided for illustrative purposes and represent a lot of informed speculation. I hope the reader finds them helpful.

Introduction

In March 1865, a 30-man team of mostly Black Union combatants traveled 100 miles behind enemy lines to conduct a ridiculously daring raid on a substantial sugar plantation in an area untouched by warfare. Then they would have to return on foot to the safety of the Union garrison in St. Augustine about 80 miles away from the remote Marion County raid site to complete the mission.

Outside of Marion County, Florida where the raid took place, it would be difficult to find anyone aware of this event, the lone military action occurring in Marion County during the Civil War. Indeed, within Marion County today, it would be difficult to find anyone aware of it. This is a retirement mecca and most have relocated here from somewhere else. Learning local history is not commonly a priority when there is golf galore.

Few know that Marion County was one of the largest slave-holding counties in Florida – Leon County, home to state capitol Tallahassee, took first place. The 1860 US Census counted 8,000 individuals in Marion County of which 5,314 were Black slaves. Since that number only counts slaves on plantations of 20 or more slaves – a peculiar census methodology unique to Florida among the southern slave-holding states – the true number was much higher, likely over 7,500, or over 75% of the actual population which was over 10,000.

For example, the 1870 Census reflects a mammoth increase in the Black population in Florida, indicating the change in census methodology to counting all individual Black people. In the 1870 census, Marion County's total population increased to 10,804, representing the White population decreasing from 3,294 to 2,926, and the Black population increasing from 5,315 (add one free Black adult in 1860) to 7,878. Of course, it would be ludicrous to imagine that Black people were flocking to Marion County, Florida after the war.

Locals tend to downplay the role of slaves in Marion's antebellum economy as 'not a big thing.' The numbers are clear: slavery was *everything* in Marion County's antebellum economy.

Many with any awareness of the Marshall Plantation raid may know of it only because of a local Civil War re-enactment held annually in early November when the snowbirds have arrived from northern climes, named "The Ocklawaha River Raid Re-enactment." The event hijacks the historical fact of the raid simply to justify a generic battle re-enactment. It gets problematic when the organizers make vague pretenses of presenting anything like the actual raid. It is farcically unlike the actual raid.

In the re-enactment, Union and Confederate soldiers, who all tend to be White, line up in uniforms with flags waving, muskets and cannons firing, and billows of smoke floating across the field of battle. Unlike the raid, this event has no Home Guard militia on horseback or Confederate cavalry, and there are no Black Union soldiers.

The Confederate soldiers in the event typically wear uniforms that Florida's Confederate soldiers did not, except perhaps officers; the soldiers wore plain, everyday civilian clothing, blending in seamlessly with the civilian population. (Confederate soldiers with uniforms were largely in the Armies of Virginia and Tennessee, but few other places.) There were no cannons in the raid.

The lone true gun battle during the raid happened in the dark moonlight somewhere in the desolate scrub between the Ocklawaha River bridge and today's Salt Springs. It involved 16-17 combatants on each side; there were no Confederate soldiers, only the local militia or Home Guard. It likely lasted only a few minutes.

Nonetheless, the re-enactment is a great spectacle and very entertaining, but *nothing* like that ever happened in Marion County. The re-enactment leaders should simply drop the pretenses of any historicity and admit that it is a generic Civil War battle re-enactment and leave it at that. People will still come to the show without pirating this actual event and corrupting/exploiting history.

The Marshall Plantation raid would be of no significance to anyone, except that in Confederate flag-flying Marion County, the raid was the only military action here during the Civil War. You can be sure it has gotten the notice of Confederate history apologists in the community. They use every occasion to celebrate local war hero Captain J. J. Dickison, an ardent Confederate, White supremacist, and local plantation owner in Orange Springs in northern Marion County with 8 slaves. Dickison played a key role in the raid narrative.

The Marion County Historical Commission, an official government body, has long been composed of Confederacy apologists with commission members generally belonging to either the Sons of Confederate Veterans or the United Daughters of the Confederacy. Thanks to these folks, some memory of the raid has been preserved, albeit in a corrupt and inaccurate form.

The Historical Commission erected an historical marker in 1999 which is set back from the roadway on County Road 314. One should forgive residents who whizz by the marker at 50+ miles per hour on a wild, forested section of the road, hardly even seeing it much less knowing what it says. The marker stands a mile or so west of the Sharpes Ferry Bridge which spans the Ocklawaha River, denoting the approximate location of the Marshall Plantation, which was north of the marker on the other side of the road.

The marker's text (see Appendix for the full text) describes the Marshall family, Col. Jehu Foster Marshall's role in the Confederate Army, his death at the Second Battle of Manassas in August 1862, and that his widow, Elizabeth Anne DeBruhl Marshall, became the plantation's owner. Not mentioned is that the Marshalls were involved in South Carolina politics and lived in South Carolina. The raid-target sugar plantation by the Ocklawaha River, as well as a cotton plantation in Wetumpka in northwestern Marion County, were income properties for the Marshalls. The Marshalls may have visited periodically, but they did not reside there; they were absentee plantation owners. It seems remarkable that the marker spends nearly half of its text talking about the Marshalls who never lived there, and the plantation itself, being a large but not exceptional operation.

Indeed, the only reason the Marshall Plantation is historically noteworthy is that it was the target of the raiders and was burned to the ground. If not for the raid, it would have been wholly forgotten along with the many other substantial Marion County plantations. However, contemporary Confederacy apologists show in the marker's text that they are interested in the plantation, not the raiders.

One more thing – there is **no** mention of slaves in the marker text, following the pattern of apologists to omit inconvenient realities like the slaves who made plantation owners wealthy. Slaves are apparently covered at the end of the marker text as "all property," lumping slaves together with wagons, mules, and horses. By the way, contrary to the marker text, "all property" was not returned to Mrs. Marshall.

Then with remarkable inaccuracy, the marker describes the accepted local version of the raid. In 1999 when the marker was installed, "Lost Cause" revisionism was still common in the social studies curriculum in Marion County public schools, celebrating the nobility and honor of what may still be called here "The War of Northern Aggression." In an inept way (or a deliberately misleading way), it seems to offer an homage to the "Lost Cause" effort at revising history to cover up the Confederacy's abominable society while also obscuring or erasing Black achievements. Since new, more accurate information about the raid has come to light, there has been scant interest in correcting the historical marker to make its description of the raid more aligned with actual history. The reader can assess the difference from what has been stated on the marker with the more informed and balanced story of the raid presented here.

We should highlight that Capt. J.J. Dickison, a key figure in the marker's text, a local fellow in Marion County, was greatly renowned and revered throughout Florida for his adventurous exploits and cunning strategic ploys as leader of Company H, 2nd Florida Cavalry, constantly plaguing Union operations across Florida. The role ascribed to Capt. Dickison in the marker text fits the expectation that Dickison heroically saves the day and brings vindication and recompense for the Union's acts of arson and thievery in its raids.

After reading the marker text, one would believe that the Marshall Plantation raid was a complete bust. The marker text leaves the impression that, all things considered, nothing of importance really happened, except the destruction of a widow's fine plantation operation and the burning of a bridge by a band of Black thieves and arsonists in Yankee uniforms. This conclusion also fits with the expectation that nothing much did happen in Marion County during the Civil War. This odd event could simply be regarded as anomalous as well as inexplicable, ultimately no more than wanton acts of violence and destruction against harmless targets.

And that is where the story stood locally until February 2016, and then the whole story was transformed.

Rick Allen's discovery

Jim Ross, an editor at the local newspaper, *The Ocala Star-Banner*, noticed a contrived kerfuffle generated by a local White plutocrat on a popular conservative-political Facebook page. The post stated that the historical marker by the site of the Marshall Plantation was suddenly missing (or someone had just noticed it was not there).

The poster's comment suggested that the marker had been stolen because it was Confederate history amid February which is Black History Month. It went on to lament the disrespect for history, racial animosity, blah, blah, blah.

Editor Jim Ross was aware of how few people even knew about the existence of the historical marker standing in the middle of nowhere. Still, it was Black History Month, and the marker was missing. He gave the slim story lead to veteran reporter Rick Allen to investigate and see what he could learn.

It didn't take Rick long to discover that the marker had likely been struck by an errant vehicle. It had been lying amid the grasses quite mangled, but it had not been stolen or vandalized. It was in the County's possession, and it would take time to replace it.

Rick considered the text on the marker. Like most local history about that era, it was primarily concerned with the Confederate-local perspective. Rick was struck by the notation about a unit of Black Union soldiers. He mused that learning about the other side of the story would be interesting and might bring this wimpy storyline to life.

He learned about the 3rd US Colored Troops that had mustered in Philadelphia, reaching out to the unit's re-enactors and historical aficionados in Philadelphia for information.

He learned about Sergeant-Major Henry James who was in his mid-30s, a "laborer" from Lancaster County when he enlisted. James was quickly advanced from a Private in Company B to be appointed Sergeant Major for the entire regiment just 12 days after his enlistment as the 3rd USCT formed its leadership ranks on the fly.

He learned that James would continue in the US Army after the Civil War as a "buffalo soldier." James contracted an infection and needed to leave the Army in 1869. He died in 1895, leaving his widow Rachel, and was buried in an AME cemetery in rural Atglen, Pennsylvania, midway between West Chester and Lancaster.

Most importantly, Rick discovered a heretofore unknown letter that appeared in the April 22, 1865 edition of the national publication of the African Methodist Episcopal (AME) church founded and based in Philadelphia, *The Christian Recorder*.

Masthead of The Christian Recorder on April 22, 1865

The Christian Recorder was widely circulated among Black Union units throughout the country following the Emancipation Proclamation and the increase in Black soldiers hungry for news and views. *The Christian Recorder* was a rarity, a Black-focused publication with broad distribution. The letter was written by Sgt. Henry S. Harmon of the 3rd US Colored Infantry, Company B whose men formed the majority of the raider team. It gives a full description of the mission and the action. (See Appendix for the full, lightly edited text of the letter.)

Rick had a big story to tell – about 2400 words – and editor Jim Ross allowed him to run with all that he had learned, publishing it on February 12, 2016 as "Mystery of the Marshall Plantation: Fallen sign, complicated past." (See link in Brief Bibliography.)

As immensely helpful as this letter was in providing a whole new perspective on the story, and as insightful as everything Rick had discovered along the way, huge basic questions remained.

Getting these questions answered requires understanding the context. The questions concern the raiders, their motivation, their goal, their reasons, and their experience as Black soldiers in the army. Considering the raid outside of the raiders' own context leaves these questions begging and leaves big issues unaddressed.

Why conduct this raid when the war is nearly over – just weeks before Lee's surrender? Why even consider a crazy raid on a target 100 miles behind enemy lines? Why do this dangerous raid when the target and its location are strategically irrelevant? Getting possible answers to these questions is our task. Critical to getting these answers is understanding the context of these Black soldiers on the raider team.

In the next few chapters, we will try to gain a better perspective on the journey of the Black soldier in the Union Army. Once that becomes clear, we can begin answering a host of pressing questions. What follows will be a summary of the Black soldiers' difficult journey.

For a wonderfully comprehensive study of the Black soldier, I highly recommend William A. Dobak's *Freedom by the Sword: The US Colored Troops, 1862-1867* which is available as a free PDF download – see the link in the Brief Bibliography. Thank you, U.S. Army Center of Military History!

Chapter One

Black Men Not Wanted

With the outbreak of hostilities following the secession of Southern states in 1861, there was a major drive to recruit men into the woefully unprepared, understaffed Union army. However, when Black men showed up at Northern recruitment centers and events, they were turned away. They were refused enlistment due to a 40-year-old policy against Black enlistment.

It made sense to recruit Black men since there was a desperate need to fill the ranks of the early-war Union army, and the war was being fought in large measure to bring a conclusion to the era of Black slavery. It would be natural for Black men to want to participate in that fight. Nonetheless, the policy remained in place.

There were huge fears about arming Black men. There was the early, disputed report about Black men being used in the Confederate army. This sparked the fear among Northerners that hordes of enslaved Black men would be deployed against Union forces.

There was also a fear among Northerners about the Union army putting arms in the hands of Black men. Arming Black men was considered a dangerous notion in the North. Racist attitudes were not confined to the South. It was one thing for Whites to profess the oppressiveness of slavery and insist that it come to an end, but it was quite another to profess racial equality and allow Black people equal standing with Whites, like authorizing Black men to carry and use firearms. For many, that was regarded as going too far.

Secretary of War Simon Cameron had to reassure Congress that neither fear was founded in truth. He had to state that there were no valid reports of armed Black slaves serving the Confederates (in fact, there was a unit of Black slaves forced to serve the Confederates in Louisiana which ended up defecting to Union forces), and the Union army would not be recruiting Black men either.

Simon Cameron

(We should note that this federal prohibition did not prevent individual states from generating militia units comprised of Black men and even led by Black officers.)

Indeed, Union army reports during the war show repeated mentions of Black people and cattle being lumped together as the same kind of thing, what was called "contraband of war."

The "contraband" term was coined by Massachusetts General Benjamin Butler amid operations at Fortress Monroe in Virginia. When a Confederate officer demanded the return of three escaped slaves, Union General Butler demurred, invoking the concept of "contraband of war," putting the escaped men to work for the Union cause at Fortress Monroe.

Gen. Benj. Butler

With the early emphasis of the conflict focused on returning secessionist states to the Union, the observance of laws requiring the return of slaves – considered property – continued to be observed by some Union officers, provided they were assured that these slaves were not and would not be used to support the Confederate effort. The ambiguity of that is self-evident since anything done by slaves in secessionist states benefited the Confederate war effort.

General Butler had had enough of the nonsense and kept the escaped slaves to work on his projects, setting the pattern for Union commanders going forward.

(Also, it was General Butler who, while commanding in Louisiana and facing far superior forces of Confederate soldiers and militia, organized escaped Black slaves into state military units under the federally approved state Governor in the Union-controlled territory, cleverly saving his strategic position and skirting the federal prohibition.)

The "contraband" concept was that property seized from the Confederates could (and should) be employed to support the Union effort. This typically meant using escaped or freed slaves for non-combat service duties like digging latrines, building fortifications, and other labor. Due to racial biases, most Union commanders never entertained the thought of arming any former slaves for combat duty.

As desperate as the Union army was to fill its ranks, even offers of willing Black soldiers were refused. War Secretary Cameron was offered 300 Black men in the spring of 1861 to assist in the defense of the threatened Capitol and refused them.

There were several independent attempts to form units of Black soldiers from populations of freed slaves in Union-held parts of Louisiana, Kansas, and Union-held areas of the Sea Islands off the Carolinas. These efforts tended to yield poorly trained, poorly equipped, and poorly led units. At least one venture collapsed when the War Department refused to pay any salary to the soldiers who were never properly recruited (or wanted) in the first place.

Although Cameron's views began changing by the end of 1861, instructing that "[if] slaves are capable of bearing arms and performing efficient military service, it is the right . . . of this Government to arm and equip them," Cameron was over-ruled by his boss, President Lincoln, who had that reference stricken from Cameron's communication.

Lincoln was not opposed to arming Black men in principle, but he was keenly aware of the delicate politics involved.

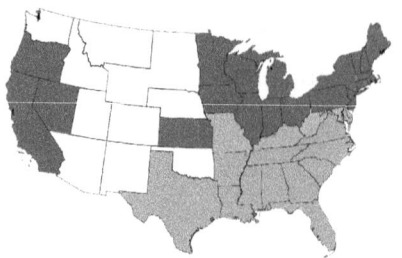

Map of border states

Key border states which had slavery present but had not seceded from the Union were of major concern for Lincoln. These were primarily Maryland, Delaware, Kentucky, and Missouri. (West Virginia formed out of secessionist slaveholding Virginia and permitted slaves after joining the Union in June 1863, not abolishing slavery until February 1865.) Lincoln knew that arming Black men at this point in the war could provide the tipping point for some or all these states to secede and join the Confederacy, making the successful prosecution of the war much more difficult. Lincoln needed time to get men and materiel together for the initially unfit Union army to become a worthy force on the battlefield.

Lincoln is said to have been confronted by a delegation of ardent abolitionists who demanded that Black men be accepted into the Union army. The delegation insisted that this fight was truly the fight for Black men, and they deserved the opportunity to participate in the struggle for freedom for their people. Lincoln responded that he had sent thousands of rifles to the men of Kentucky to serve the Union war effort. Lincoln pointed out that arming Black men in the Union army could have those thousands of rifles turned against Union forces. That was the end of the discussion.

By July 1862, Lincoln was able to tell his Cabinet of his plans to issue the Emancipation Proclamation. A carefully worded document that avoided a wholesale freeing of all slaves in all states, the Proclamation applied to states that had seceded – not the border states – and to territory taken from those secessionist states, where slaves were to be freed and extended the opportunity to serve the Union army. In short, slavery would be allowed to continue in many places, even those controlled by the Union military, thanks to the series of exceptions listed in the Proclamation.

Yet the Proclamation would remain unpublished. There was something else Lincoln needed in order to proceed which was critical in signaling to the border states that their best choice was to remain in the Union.

Desperately lacking in the summer of 1862, over a year into the war, was a real victory by the Union army. Union forces had yet to show that they could beat the Confederate forces. Lincoln needed that big victory to demonstrate to the border states that the Union was going to win this war.

Commander of the Army of the Potomac General George B. McClellan had built an impressive, well-equipped force. He was tasked with taking on General Joseph E. Johnston's Confederate army in Virginia with the goal of taking Richmond ultimately. The dashing McClellan was proud of his army, and his men were deeply devoted to him. However, like all of Lincoln's leading generals throughout the war, he had issues.

McClellan adored his army but was loath to get it messed up and possibly ruined by things like warfare and combat. He tepidly pursued Johnston's army, engaged in combat, and when he saw losses coming from the battlefield, he would disengage, leaving Johnston's army to regroup. He could never muster the fortitude to prosecute much of a battle, allowing his foe to continue to play "tag" and await the next round further up the road. Johnston proved to be just as cautious as McClellan.

McClellan was also prone to grossly overestimating the size of the force he faced, sometimes believing it to be exponentially larger than it actually was. (McClellan's army always outnumbered his Confederate foe.) Believing that his whole army was in danger of annihilation from superior forces, he would fail to press an attack. This was the essence of the ill-fated Peninsula campaign.

When Johnston gets wounded, General Robert E. Lee leaves his command in the southernmost states, including Florida, and takes over the Army of Virginia, deterring McClellan's force from moving against Richmond.

McClellan then chases Lee across the Virginia countryside all the way to Sharpsburg, Maryland by Antietam Creek in September 1862. Here the opposing armies are ready for a full-fledged confrontation.

Lincoln does not want another one of McClellan's cautious, indecisive failures; he needs and wants a victory as the Emancipation Proclamation silently languishes without that victory. Famously, Lincoln leaves Washington, DC to travel to Sharpsburg and confer with McClellan in the field to ensure that his general takes the fight to his enemy and prevails.

Lincoln meets McClellan

At Antietam, an ugly slaughter ensued as wave after wave of Union units charged into a steady hail of Confederate gunfire. The casualties were staggering for the Union forces, but they were nearly as horrific for the Confederate forces with killed, wounded, and captured exceeding 20,000 total for both sides. It is said that Lee retreated because his soldiers were running out of ammunition to kill Union attackers. McClellan was faulted yet again for failing to pursue the badly debilitated army of General Lee, leading to McClellan's eventual removal from command several months later.

In any case, as horrendous as it was, Antietam was the Union victory that Lincoln sought, having successfully driven Lee's army back across the Potomac River into Virginia. Five days later, on September 22, 1862, Lincoln announced the Emancipation Proclamation to take effect on January 1, 1863.

Chapter Two

The Black Soldier "Experiment" Begins

Congress had passed the Militia Act in July 1862 which enabled the president to organize and utilize "persons of African descent" for "the service of the United States, for the purpose of constructing intrenchments, or performing camp service, or . . . any military or naval service for which they may be found competent." This Act also set the pay rate at $10 per month, the same as laborers at Fortress Monroe under General Butler and the Navy sailor recruits (recruitment of Black sailors was authorized by Navy Secretary Gideon Welles in September 1862), but less than the $13 per month paid to White army privates. Further, there was a $3 deduction for the uniform each month, leaving a Black soldier with just $7 per month pay.

Recruitment efforts for Black men were formally authorized by the Emancipation Proclamation which took effect on January 1, 1863. The 1st Kansas Colored Volunteer Infantry – a persistent renegade Black recruitment effort first organized as a Kansas state militia unit – was finally recognized by the US War Department in January

1863, and then the 54th Massachusetts Infantry became organized in March, the unit which became well-known to contemporary generations thanks to the 1989 movie *Glory*. Lincoln became more outspoken in supporting the recruitment of Black soldiers.

However, the process was complete chaos. The disparity in pay between White soldiers at $13 per month and Black soldiers at $10 per month noted above was not corrected until early 1864! It led the 54th Massachusetts (and then the 55th Massachusetts) to famously refuse their pay altogether until it was on par with White soldiers.

In November 1863, Massachusetts Gov. Andrew signed an act that would have the state make up the White-Black federal pay difference to bring parity. One soldier's letter was quoted by a writer in the *Boston Journal* pointing out the Governor's perhaps well-intended but mistaken notion that this pay refusal was about the sum of money as opposed to a matter of principle based on equality - *[the Governor's action] in effect, advertises us to the world as holding out for money and not from principle — that we sink our manhood in consideration of a few more dollars.*

Even when Congress finally addressed the issue in early 1864, the pay rate for Black soldiers only applied to those who were free prior to the war's commencement on April 19, 1861. For escaped slaves who were enlisted into the Army, the pay rate remained unchanged. There was a declaration to be made by Black soldiers that they were indeed free from April 1861, but it was never checked. This led many who were not free to affirm nonetheless that they were free in order to qualify for the pay upgrade. The pay-refusing 54th and 55th Massachusetts soldiers were once again insulted by the stupid rule imposed by Congress and continued to refuse their pay on these terms.

Units were formed based on geographic locale. Black units needed to have White officers appointed for their command since Black soldiers were not permitted to be commissioned officers like Lieutenant, Captain, Major, or Colonel. The thought of a superior-ranking Black man commanding a White man was hardly plausible at this time. Again, racist beliefs were common among Northern states, their governments, and their citizens.

Local officials were often in charge of appointing White officers to Black units. However, the appointments made by local officials could easily conflict with the appointments made by the unit's commanding officer. Who decided who was in charge? There were constant appeals being made to higher-ranking commands for one officer over another.

Letter writer Sgt. Harmon's pay record for September to October 1864 noted in the "Remarks" that he was "Free on or before April 19th 1861"

This confusion also affected the War Department which may not have known where an officer was assigned in order to pay them. One such officer was still seeking back pay for six months of service from the Civil War in 1884!

Some of these officers were adept at fraud, getting their Black soldiers to entrust their pay to them for safekeeping. When sufficient funds had been collected and before difficult questions began to be asked, these officers would slip away with their ill-gotten gains.

One should not imagine that there was anything like the contemporary military standards, training, and experience for commissioned-officers. An appointee may be assessed as warranting an appointment to a certain rank based on their physical height or how well they knew a local politician rather than having any worthwhile leadership skill, military experience, or even desire to work with Black soldiers. There was rank hypocrisy among some appointees; some of those who were most negative about Black men being soldiers were the most eager for an appointment to lead a Black soldier unit.

As the war progressed, effective commanders were discerned as well as ineffective ones. The effective commanders tended to be advanced in key positions while the ineffective ones would be relegated to assignments unlikely to be of real consequence. This often meant ineffective officers being assigned to command Black units in less consequential places like the Department of the South or the Department of the Gulf.

Black soldiers were also subject to violence from their White counterparts. The practice of Black soldiers carrying concealed weapons for self-protection was so significant that orders had to be issued to prohibit the practice.

Simply because the Union army was accepting Black soldiers into its ranks did not mean that they were respected at all. Racial bias commonly defined Black soldiers as unworthy, believing that they would not stand in combat and would flee under fire.

"Fatigue" duty for Black soldiers after the Cold Harbor battle

Black soldiers were commonly given the hard labor tasks that White soldiers (any soldier) disdained. These were called "fatigue" assignments, tasks like constructing earthworks, digging latrines and trenches, and burying the dead. Commanders had to issue orders insisting that White soldiers were the same as Black soldiers and that all duties, including fatigue assignments, were to be shared equally.

Units from which the Marshall Plantation raiders were drawn

In August 1863, the 3rd US Colored Infantry (3rd USCT) was mustered in Philadelphia, having trained at Fort William Penn (ironic that a pacifist Quaker's name is attached to a military installation). Naturally, the 3rd USCT drew enlistments from eastern Pennsylvania, men like Henry James of Lancaster County, west of Philadelphia, a 35-year-old who listed his occupation as a laborer. James enlisted on June 30, 1863, but received starting pay of just $7 per month, comprising the standard rate for Black soldiers of $10 per month less $3 per month for the cost of the uniform. James went from Private to an appointment as Sergeant-Major of the regiment just 12 days later. This made him the highest-ranking Black (non-commissioned) officer in the 3rd USCT, an indication of the leadership qualities he must have displayed in the newly forming unit.

The 3rd USCT had a stunning battle flag created by Black artist David Bustill Bowser who was commissioned to create a number of battle flags for the newly forming Black regiments. Shown here, the 3rd USCT battle flag presents a Black Sergeant jointly holding the national flag with Columbia, the feminine symbol of the republic, with encampment tents in the background. The top ribbon states: *Rather die freemen, than live to be slaves*, and the name of the unit, 3rd United States Colored Troops, is on the bottom ribbon. (Bowser used a similar motif for the 127th USCT with a Black soldier with a hand raised up to the national flag being carried by Columbia also with encampment tents in the background, but with the 127th's motto on the top ribbon: *We will prove ourselves men*.)

3rd USCT battle flag

The 3rd USCT would be led by Colonel Benjamin Chew Tilghman. He had volunteered at the outbreak of the war and was a Captain in the 26th Pennsylvania Infantry. He would be given the rank of Colonel and command of the 29th Pennsylvania Infantry. He was badly wounded in the thigh at the battle of Chancellorsville in the spring of 1863. After

his recovery, he would return to accept command of the 3rd USCT. (After the war, Tilghman would become known for inventing and patenting the process of sandblasting, based on an insight gleaned from his time in military service.)

Tilghman would complain that his regiment had been poorly trained and was not ready for action after just a few weeks of training, but the 3rd USCT was assigned to the Department of the South and sent to Morris Island, South Carolina in August 1863 for the campaign there concerning Forts Wagner and Gregg at the mouth of Charleston Harbor. More on that campaign later.

The Union Army's Department of the South encompassed operations in South Carolina, Georgia, and Florida. (The Florida Panhandle west of the Apalachicola River would be assigned to the Department of the Gulf in August 1862, and later Key West and the Dry Tortugas as well in March 1863.)

The Department of the South would establish its main base on Hilton Head Island, South Carolina which is located between Savannah to its south and Charleston to its north, near Beaufort. Hilton Head Island was seized in a massive 12,000-man amphibious Union landing in a joint Army and Navy operation in late 1861, the largest amphibious landing ever made to that point. From Hilton Head Island, Union forces engaged in campaigns along the southern Atlantic coast, often working in concert with the Navy and its blockade efforts. A number of forts remained in Union possession after secession, dotting the Gulf and Atlantic coastlines assisting with blockade efforts, and acting as bases for launching raids and other operations into the interior.

While Black units formed in the North among free men, there were also Black units formed in the South following the January 1863 Emancipation Proclamation. Drawn from freed and escaped slaves, these units included the 1st South Carolina Volunteer Infantry (1st SCVI), the 2nd South Carolina Volunteer Infantry (2nd SCVI), and eventually the 1st North Carolina Colored Volunteers (1st NCCV). Soldiers from the 2nd SCVI (later renamed the 34th US Colored Troops) would be included among the raiders.

As noted earlier, there were independent efforts to organize former slaves into military formations, and some of the earliest units authorized in 1863 resulted from those previously deterred attempts, like the 1st Kansas Colored Infantry which had originally been organized as a state militia unit before becoming the first unit of Black soldiers authorized as a US Army unit. Having been formed largely from former slaves, the 1st KCI had already proven themselves in action in October 1862 at the Battle of Island Mound, Missouri.

Capt. William Mathews

(It should be noted that much credit for recruiting men for the Kansas unit went to William Mathews, a free Black businessman and a station master for the Underground Railroad. Mathews held the rank of Captain for the state-authorized unit until the unit received its War Department authorization, at which time he was demoted from that commissioned officer status. Commissioned officer status would largely not be available to Black soldiers during the war, except for a precious few who were commissioned late in the war.)

Recruitment efforts on the Union-seized Hilton Head Island, South Carolina began in the summer of 1862 under Union Major General David Hunter who was commander of the Dept. of the South. Boldly, he declared all slaves in South Carolina, Georgia, and Florida free in May 1862, long before Lincoln's Emancipation Proclamation. Then, to accelerate recruitment for Black units, Hunter ordered all free Black men of the Union-occupied Sea Islands of South Carolina between 18 and 45 years old "capable of bearing arms" to report for military training – forced conscription!

Gen. David Hunter

It did not go well. Lincoln demanded Hunter's conscription order be rescinded but he wouldn't instruct Hunter to disband the military unit. Still, the War Department would not support the unit which, by the way, sported snappy crimson-red pants with blue coats. The unit was mostly disbanded in August 1862. However, by October, the War Department had changed its mind and, eventually, the 1st South Carolina Volunteer Infantry was formed officially on January 31, 1863. It would be headed by an abolitionist Unitarian minister from Massachusetts, Colonel Thomas Wentworth Higginson.

The 2nd South Carolina Volunteer Infantry came about when Colonel James Montgomery of Kansas was authorized by the War Department to organize units of former slaves from Georgia and Florida on Hilton Head Island, SC.

Montgomery had built a reputation for himself in the violent conflicts in late 1850s Kansas amid the struggle over whether the territory would enter the Union as a free state or a slave state. A vehement abolitionist, Montgomery enthusiastically embraced a guerilla-style strategy named "jayhawker," a moniker since adapted generally for Kansans and by the University of Kansas. "Jayhawker" operations in the "bloody Kansas" conflict involved targeting pro-slavery sites, businesses, and individuals with acts of vandalism, arson, and violence as well as for the profit and benefit of the anti-slavery attackers. Montgomery would continue to employ this "jayhawker" strategy with the 2nd SCVI. Again, soldiers from the 2nd SCVI would also be part of the raider team.

James Montgomery in 1858

It is interesting to note that while Montgomery was passionately opposed to slavery, he was nonetheless a racist. A passionate harangue came from Montgomery on September 30, 1863 on Morris Island, SC when there was a renewed refusal by the Massachusetts regiments to accept payment on unequal terms compared to White soldiers. He is essentially recorded by a witness[1] as saying:

> *Men: the paymaster is here to pay you. You must remember you have not proved yourselves soldiers. You must take notice that the Government has virtually paid you a thousand dollars apiece for setting you free. Nor should you expect to be placed on the same footing with white men. Any one listening to your shouting and singing can see how grotesquely ignorant you are. I am your friend and the friend of the negro. I was the first person in the country to employ nigger soldiers in the United States Army. I was out in Kansas. I was short of men. I had a lot of niggers and a lot of mules; and you know a nigger and a mule go very well together. I therefore enlisted the niggers, and made teamsters of them. In refusing to take the pay offered you, and what you are only legally entitled to, you are guilty of insubordination and mutiny, and can be tried and shot by court-martial.*

---

1. Emilio, Captain Luis F.. *A Brave Black Regiment: The History of the Fifty-Fourth Regiment of Massachusetts Volunteer Infantry 1863-1865*, p. 68

Chapter Three

Black Soldiers in Action in the South

The 2nd SCVI was mustered into service in May 1863 and for a time was paired in a brigade with the 54th Massachusetts and Colonel Robert Gould Shaw, engaging in operations along the south Atlantic coast. (Again, the 54th Massachusetts was featured in the 1989 movie *Glory*.)

The 2nd SCVI was involved in the early June raid conducted in a joint Army and Navy operation together with Harriet Tubman up the Combahee River in South Carolina. Using Tubman's network among the region's slave population, the slaves on plantations along the river must have been alerted to come to the river when the smoke billowing from Union gunships and transports was visible, or when the gunships' cannons opened fire on the plantations. Incredibly, roughly 800 slaves escaped to the river, overwhelming any efforts

Harriet Tubman

by overseers to detain or deter them, but also overwhelming the expectations of the Navy which soon saw its ships filled to their capacity with escaping slaves. William C. Heyward, the owner of the Cypress Plantation, listed the plantation's losses at 199 slaves and the destruction of over $50,000 in buildings, thanks to the work of Navy gunboats and the 2nd SCVI deployed in "jayhawker" operations.

Raids by the 2nd SCVI often followed Col. Montgomery's jayhawker method. Such harsh tactics went to their logical conclusion with Montgomery's command to burn down the entire town of Darien, Georgia on June 11, 1863,

Combahee River raid

simply because the town had supported the Confederacy. The wanton destruction had no military purpose and only served to harden the prevailing negative attitude toward Union soldiers as vicious invaders. The 54th Massachusetts under Col. Shaw was in a brigade with the 2nd SCVI with Montgomery being the brigade's commanding officer. Col. Shaw and the 54th Massachusetts were caught up in the operation at Darien. Shaw would later complain to Montgomery about the unnecessary and unwarranted destruction leveled against the town. Montgomery remained unyielding, convinced that it was entirely appropriate.

The brigade combining the 2nd SCVI and the 54th Mass. was also deployed to the Morris Island campaign off Charleston, South Carolina in July 1863. They were dispatched to James Island where their camp was attacked by Confederates at Grimball's Landing on July

16. It was the first battlefield engagement for the 54th Mass. The attack failed with the Confederates driven off, and then the Union brigade was withdrawn. That move on James Island was simply one of two feints by Major General Quincy Gillmore to distract Confederate forces from the main thrust which would be a renewed attack on Fort Wagner.

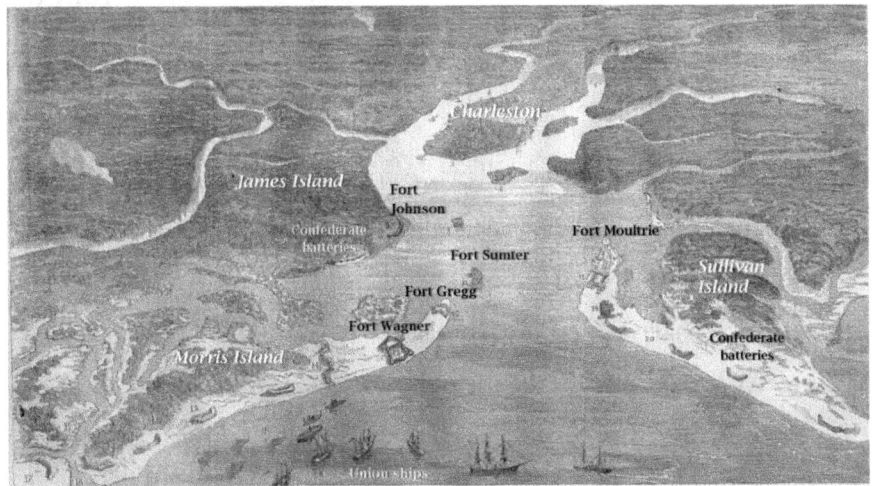

Charleston Harbor fortifications

As the map shows, the Union's desire to capture Charleston would be hard fought, requiring the capture of the numerous forts dotting the heavily fortified harbor. Forts Wagner and Gregg on Morris Island needed to be secured first. These would be formidable undertakings.

Fort Wagner - Left: inside view; Right: beach-side view.

Fort Wagner had thick, palmetto log-reinforced sand walls that stood at places a daunting 30 feet above the beach. Outside the walls, an attacker would find a water-filled trench, land mines, and sharpened palmetto stakes. The narrow island was made narrower by a heavy swamp covering half of Morris Island's width in front of the fort, leaving only a thin ribbon of beach leading to the high fort walls which had an unobstructed view. A beachfront assault from the water was completely suicidal, but a land-based assault along the ridiculously narrow strip of beach was nearly as bad.

Major General Gillmore had made his name by using long-range rifled cannons in subduing Fort Pulaski on Cockspur Island amid the Savannah River in Georgia. Fort Pulaski had brick and masonry walls 11 feet thick that seemed impenetrable for standard Union artillery. Gillmore's introduction of longer-range rifled cannons brought impressively devastating results, leading to Fort Pulaski's surrender.

However, Fort Wagner's larger and thicker reinforced sand and earthen walls would prove to be a different kind of target from the hard brick and masonry walls of Fort Pulaski in Georgia.

The First Battle of Fort Wagner resulted from General George C. Strong's brigade having landed at the southern end of Morris Island and then moving north to engage Fort Wagner. The assault at dawn amid heavy fog on July 11, 1863 was disastrous, resulting in 339 Union casualties while the Confederate defenders had only 12 casualties in repulsing the attack. This failed attempt should have been a sufficient lesson of the folly of any direct assault on Fort Wagner, but it wasn't.

Brigadier General Truman Seymour, an 1846 West Point graduate who had seen plenty of action in other theaters during the war, was given command of the Morris Island ground forces to launch another attack on Fort Wagner a week after the first failure. In planning the attack, Seymour is quoted by one observer as having said, "Well, I guess we will . . . put those damned niggers from Massachusetts in the advance; we may as well get rid of them, one time as another." We will see other instances of General Seymour's racist disdain for Black soldiers.

Major General Gillmore had the Navy conduct a close-range 16-hour bombardment of Fort Wagner prior to the second assault. The reinforced sand composition of the fort's high, thick walls mostly absorbed the impact of the Navy shells, having a negligible impact on the massive fortifications.

Nonetheless, Seymour would proceed with the second attack, placing the 54th Massachusetts in the lead position among the forces arrayed. (The 2nd SCVI was on Morris Island but not part of this second assault.)

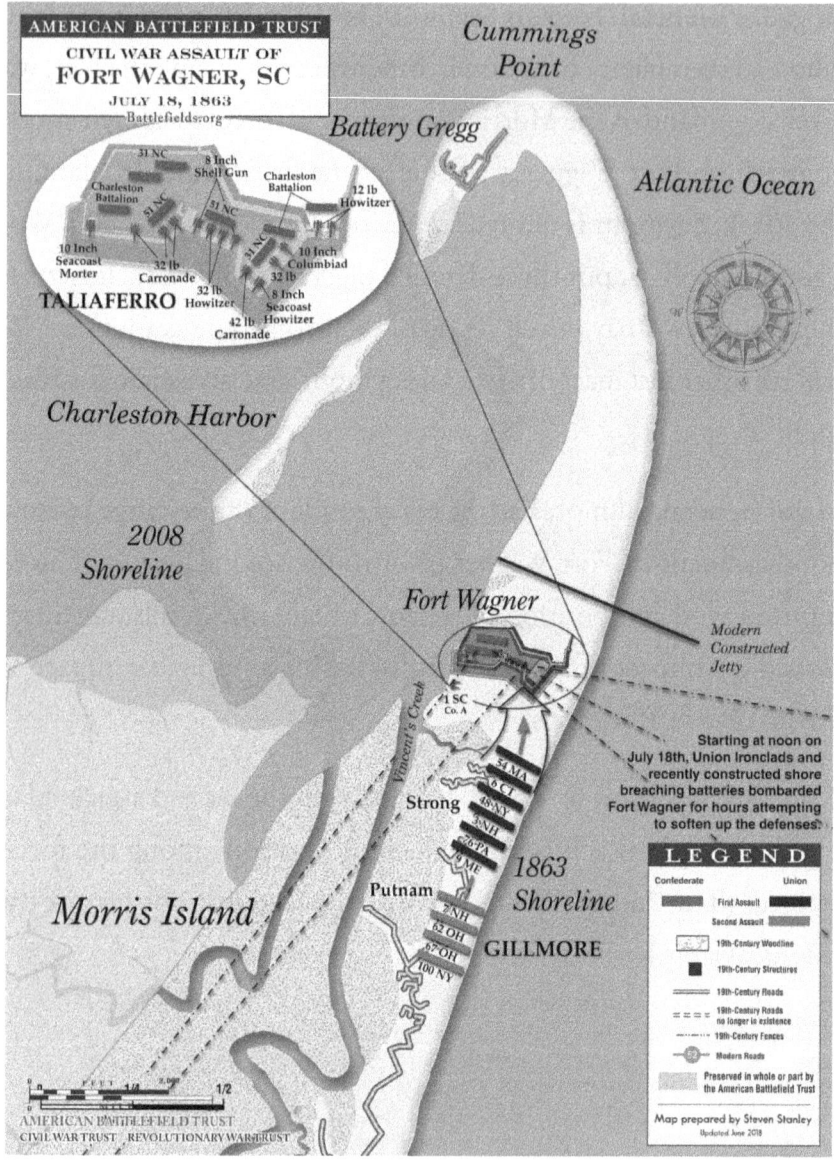

Fort Wagner battle map

As portrayed so vividly in the 1989 movie *Glory*, the attack begins in the dark at 7:45 pm on July 18. The 54th Massachusetts would charge to within a few hundred yards of the battlements when the Confederates opened their murderous fire with cannons and rifles. The barrage decimates the 54th. Col. Shaw is killed atop a high parapet in rallying his men further forward. The 54th will get to the relative safety of the base of the wall, scale it, and engage in hand-to-hand combat with the fort's defenders but they are unable to proceed. By 10 pm, the awful battle was over as Union forces withdrew.

Union forces suffered over 1500 casualties with several hundred lost from the 54th Massachusetts alone which was reduced to just 315 men. Besides Col. Shaw, there were fatal injuries sustained by General George C. Strong who had led the first assault a week before, plus two other colonels. Confederate casualties numbered only 174.

Col. Robert G. Shaw

While the assault on Fort Wagner was a failure, the incredible bravery and determination of the 54th Mass. would convert the cynical who were sure that Black soldiers were racially unsuited to the tasks of combat and believed that they would turn tail and run when the going became tough. After Fort Wagner, such cynical comments were thoroughly undone and praise began to be heaped on the remarkable courage and heroic fighting spirit displayed by the 54th Mass. in that ghastly assault.

There would be no further attempts at a frontal assault on Fort Wagner. It was placed under siege and bombarded regularly by artillery. The 2nd SCVI along with the 3rd US Colored Infantry, arriving from Philadelphia in mid-August, would take part in the siege operations. This meant pursuing General Gillmore's new strategy of digging trenches through the sand to gain proximity to the walls in order to breach them. Daytime work was too hazardous with units coming under fire. Work turned to nighttime operations. Still, these were the summer months in South Carolina, laboring under enemy gunfire amid the harsh heat, humidity, and regular rains as they built fortifications and dug trenches on a sea island, typically working in mud and standing in water up to their knees.

Over the roughly two-month-long siege of Fort Wagner, an incredible amount of "fatigue" labor was involved, disproportionately borne by Black units. In *A Brave Black Regiment*, the 54th Massachusetts Captain Luis Emilio states:

Three-quarters of all the work was at night, and nine-tenths under artillery and sharpshooters' fire or both combined. Regarding colored troops, Major Brooks, Assistant Engineer, in his report, says, — 'It is probable that in no military operations of the war have negro troops done so large a proportion, and so important and hazardous fatigue duty, as in the siege operations on the island.' The colored regiments participating were the Fifty-fourth and Fifty-fifth Massachusetts, First North Carolina, Second South Carolina, and Third United States Colored Troops.

By the time Union soldiers had gotten their trenches to the walls to breach them in September, Confederate defenders had spiked their cannons and abandoned the forts.

The 2nd SCVI and the 3rd USCT would have roles to play in ongoing operations around Charleston for the remainder of 1863.

Chapter Four

Black Soldiers in Action in Florida

In February 1864, Major General Gillmore decided it was time to occupy Jacksonville, Florida for the fourth (and final) time during the war. Among the units dispatched to Jacksonville are the 2nd SCVI, the 54th Massachusetts, and the 3rd USCT.

At the same time in February 1864, the three southeastern units composed mostly of former slaves were all reorganized and re-titled. The 1st SCVI becomes the 33rd USCT, the 2nd SCVI becomes the 34th USCT, and the 1st NCCV becomes the 35th USCT.

Gen. Quincy A. Gillmore

Gillmore appointed General Truman Seymour, the same general who commanded the ill-fated Second Battle of Fort Wagner, as commander of the District of Florida. Fifteen regiments of Union soldiers (7 of Black soldiers), about 5500 men for starters including the 3rd USCT and the newly labeled 34th USCT (former 2nd SCVI), easily occupied nearly deserted Jacksonville.

In a mutual agreement between the US Navy and Confederate leaders, Jacksonville had been occupied but not fortified by Confederate soldiers as US Navy gunships had kept watch and control of the two rivers, the St. Johns River (including Mayport and Jacksonville) which flows from the south in the middle of the state northward, and the St. Marys River (including Fernandina) which runs west to east along the border between Florida and Georgia.

As in the previous occupations of Jacksonville, there was little real resistance from the small number of Confederate soldiers present who cleared out fast.

Most residents had left town during previous occupations, either as Union sympathizers being evacuated with Union troops or as Confederate sympathizers fleeing with their goods and slaves into more favorable sections of the interior.

Images of Jacksonville, Florida in 1865

General Seymour arrived in Jacksonville two days later on February 9, 1864, and the 3rd USCT was sent to occupy Baldwin, a railroad depot (still the site of a railroad depot today).

Seymour sent a message to his superior, Major General Gillmore, who had been to visit him in Jacksonville and Baldwin. In his letter, Seymour decried the effort to return Florida to the Union as a "delusion," and urged the withdrawal of forces to Jacksonville in order to hold the town, believing that far more men would be needed for any operations into the interior.

Gen. Truman Seymour

However, he also stated that Palatka, one of the few significant settlements on the St. Johns River, should also be held, never acknowledging that Palatka was over 60 miles south of Jacksonville, i.e. deep in the interior!

Heading back to his South Carolina headquarters, Gillmore replied that Seymour should continue the limited operations already engaged at Baldwin and on the St. Mary's River, perhaps including Palatka if that was indeed possible, and to confer with him before acting.

Also, part of the Jacksonville occupation force was President Lincoln's personal secretary John Hay who was there to help organize Union sympathizers to bring about a return of Florida to the union. According to Hay:

John Hay, 1862

> *Seymour has seemed very unsteady and queer since the beginning of the campaign. He has been subject to violent alternations of timidity & rashness now declaring Florida loyalty was all bosh—now lauding it as the purest article extant, now insisting that* [General Pierre G.T.] *Beauregard was in his front with the whole Confederacy & now asserting that he could whip all the rebels in Florida with a good Brigade.*

This completely inexplicable and confounding reversal of attitudes was about to become a huge problem.

Meanwhile, Confederate General Beauregard was alarmed at the size of the Union forces now present in northeast Florida. The much-needed supply lines from Florida's breadbasket could quickly become imperiled if Union forces moved west toward Tallahassee.

Gen. P. T. Beauregard

The previous summer, Union General U. S. Grant had prevailed in his siege of Vicksburg, Mississippi, taking dominant control of the southern Mississippi River, and generating logistics and resupply woes for the Confederate forces. Confederate forces could not afford to lose their supply line from Florida.

Beauregard knew that General Finegan, who commanded the scattered Confederate forces in north Florida, was stretched thin before the new Union presence, having at best 2000 soldiers under his command. Finegan at this time would hardly be able to stem any Union advance. Beauregard dispatched over 3000 reinforcements from Georgia to Finegan.

Battle of Olustee

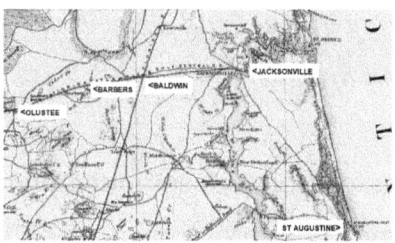

Believing that the Confederate forces were minimal based on reports of Union skirmishing encounters with small units and finding thin resistance, Seymour had another sudden, bizarre change of attitude.

On February 17, Seymour announces his plan to take 5500 men (the total Union force under Seymour's command grew to about 9000 men) and move west, intent on taking the key railroad bridge over the Suwanee River, over 100 miles west of Jacksonville!

The Union forces departed, following the railroad right-of-way west from Jacksonville for several days, traveling nearly 50 miles. This brought the force to the vicinity of Olustee Station and Ocean Pond, about 12 miles east of Lake City. The Union approach was hardly a secret and preparations were made to meet the Union force as the Confederate reinforcements from Georgia were rushed into position around Olustee.

Meanwhile, Major General Gillmore arrived at his headquarters at Hilton Head Island and received a communication from Seymour declaring his plans for a major expedition westward, plans which were *already* underway. Seymour also asked for a diversionary move in Georgia in concert with Navy gunships to keep the Confederates from sending reinforcements. Gillmore is barely able to check his incredulity in reply:

> *I am just in receipt of your two letters of the 16th and one of the 17th, and am very much surprised at the tone of the latter and the character of your plans as therein stated. You say that by the time your letter of the 17th should reach these headquarters your forces would be in motion beyond Barber's moving toward the Suwannee River, and that you shall rely on my making a display upon the Savannah River, with "naval forces, transports, sailing vessels," and with iron-clads up from Wassaw, &c., as a demonstration in your favor, which you look upon "as of great importance." All this is upon the presumption that the demonstration can and will be made; although contingent not only upon my power and disposition to do so, but upon the consent of Admiral Dahlgren, with whom I cannot communicate in less than two days. You*

must have forgotten my last instructions, which were for the present to hold Baldwin and the Saint Mary's South Fork, as your outposts to the westward of Jacksonville, and to occupy Palatka, Magnolia, on the Saint John's. **Your project distinctly and avowedly ignores these operations** [emphasis added] *and substitutes a plan which not only involves your command in a distant movement, without provisions, far beyond a point from which you once withdrew on account of precisely the same necessity, but presupposes a simultaneous demonstration of "great importance" to you elsewhere, over which you have no control, and which requires the co-operation of the navy.* **It is impossible for me to determine what your views are with respect to Florida matters** [emphasis added] ....

There is much more to this correspondence, but the reader gets a solid taste in the excerpt above for Gillmore's dismay and disdain for Seymour's rash and irresponsible action.

At Olustee, the Confederates had chosen a pine forest for concealment, adding some hastily dug trenches to secure the position further. A skirmishing unit was sent out to draw in the approaching Union soldiers.

On February 20, the forces met. One of the Union units in the forward position was the 8th USCT which had mustered in Philadelphia only in November and had received poor training with no combat preparation or experience. Indeed, their commander Colonel Charles W. Fribley had authorized soldiers to shoot their muskets after sentry duty so that they would get the feel of handling their primary firearm. This hardly prepared them to be among the frontline units at Olustee in what even veteran Union officers called one of the most intense assaults they had ever experienced.

Placed among the leading units, the 8th USCT was initially terrified, unable to return fire amid the hail of bullets and cannon fire as men dropped all around them. They had no real cover and suffered horrific losses with more than 50 percent casualties in just three hours. Col. Fribley was killed in action. They would recover to begin engaging, but the 8th USCT was far from the only Union problem at Olustee.

Seymour, apparently failing to believe the size of the Confederate force initially, did not deploy his units together, but sent in units one after another only to have them decimated by enemy fire and then replaced with more. The veteran 54th Massachusetts (again) and the 35th USCT (1st NCCV) under Col. James Montgomery – noted earlier of "jayhawker" notoriety – were brought in to stop the Confederate advance and provide cover for the 8th USCT and other units withdrawing from the field after spending hours in intense combat. For the 35th USCT, this would be their first real combat engagement, too, and they would suffer the loss of several hundred men. On another flank, Col. Barton had three White New York regiments engaged in a firefight for four hours, incurring horrific casualties to 800 men – over 300 killed – including all three regimental commanders.

Colonel Montgomery called for a retreat as men began running low on ammunition, the sun was setting, and the fight had proven not only fruitless but also far too costly. Confederate forces would continue to advance as the Union units withdrew up to a stream before their rapid retreat dissolved into a chaotic flight, covered still by the 54th Massachusetts joined by the 7th Connecticut Infantry and a cavalry unit.

The retreat of Union forces became complete mayhem as the wounded were left behind along with packs and gear, guns and ammunition, and other stores amid the panicked run eastward. By midnight, Union forces had reached Barber's Station (now MacClenny) where they had begun the day about 15 miles from the battlefield, over 30 miles west of Jacksonville.

Confederate command chose not to pursue the raggedly and rapidly retreating Union forces, leading to harsh criticism later. The lethargic advance of Confederate General Finegan allowed the crippled, disorganized Union force to make its way back to Jacksonville. It isn't until the 22nd – two days after the battle – when Confederate forces arrive in Union-abandoned Baldwin.

Gen. Joseph Finegan

Indeed, on February 22, Union forces were still coping with problems that delayed their messy retreat. The lone available locomotive for the train bearing the many wounded broke down by Ten Mile Station, as the name suggests roughly 10 miles from Jacksonville. It was the only locomotive available; in resource-stripped northeastern Florida, there was no replacement.

Again, the duty falls to the 54th Mass. to address the problem. Having covered the Union retreat from Olustee and prevented a massacre, and then marching all the way to Baldwin, the exhausted 54th is ordered to turn around and march back to Ten Mile Station and move the train loaded with the wounded. Using ropes, the unit pulls the entire train three miles before horses are added for the rest of the journey to Jacksonville, a total of 9 miles in an effort that took 42 hours overall to complete. [1]

Over 850 wounded were taken back from the battlefield by Union forces, but dozens were left behind. Confederate reports indicated clearly (and rather delightedly) that a number of the Black soldiers left on the battlefield were summarily shot and killed, an all-too-common practice among Confederates in dealing with Black soldiers wounded or taken prisoner.

While Union forces had greater casualties – over 1850 men altogether – making the Battle of Olustee proportionately the second worst

---

1. The manual movement of the train bearing the wounded by the 54th Mass. is rarely mentioned. It is noted on the battle's Wikipedia page but rarely elsewhere. *A Brave Black Regiment* was compiled by Luis Emilio, a Captain with the 54th Mass. throughout the war. His 1891 volume details this exceptional action in being recalled from Baldwin and manually moving the train by the 54th Mass. on p. 84.

battle in the war for Union casualties at about 34 percent of those deployed. The Confederate forces took heavy losses as well – nearly 950 casualties at about 19 percent of those deployed.

Of the units that we have been following, the 3rd USCT was holding Baldwin during the Olustee fiasco and then ordered to withdraw. The 34th USCT (2nd SCVI) remained behind in Jacksonville. With many of the men of the 34th USCT having been slaves who escaped captivity in Florida, and with both the 34th USCT and 3rd USCT having been seasoned by combat action already, why Seymour left them behind and brought along the poorly trained, inexperienced, Philadelphia-area men of the 8th USCT, as well as the equally green 35th USCT, is simply another head-scratcher for anyone trying to understand how an operation could be so ineptly mismanaged by a veteran senior officer.

After Olustee

General Seymour would be transferred to a new assignment in Virginia by the end of March 1864. Seymour would then be captured during the Battle of the Wilderness in May, leading one of his harshest critics, Major John W. M. Appleton of the 54th Massachusetts, to state frankly: "They are welcome to him."

Major General Gillmore was asked by Major General Halleck in February how many men could be spared for other operations. Gillmore thought relocating 7,000-11,000 would leave the Dept. of the South with enough for maintaining its defense.

Then in April, Gillmore would get tapped by Lieutenant General Ulysses S. Grant for command of the newly forming X Corps.

For his new assignment, Gillmore took with him 40% of the Dept. of the South's total manpower compared to the one-third expected previously. With the priority of the ongoing operations to take Charleston, the drastic reduction in troop strength meant that there would be no large-scale campaigns like Olustee in Florida again, however, smaller-scale operations would continue.

While the Union forces would make extensive use of the St. Johns River to carry out its operations, there was never enough manpower to sustain more than a temporary occupation or an extended incursion. Working in concert with Navy gunships and transports, the Union found the St. Johns River to be an excellent conveyance to move strong numbers of soldiers into the interior for extended operations. When missions were completed, the gunships and transports provided the means of reasonably safe extraction back to Palatka, if it was being occupied, or else to Jacksonville.

After Olustee, the Confederate leaders realized that securing Florida and its valuable supply lines needed to be a priority. With such a large and growing Union presence in northeast Florida, Confederate General Beauregard increased the troop strength to 8,000 with plans to increase it much further. The Confederate effort could hardly afford to do this since there was a desperate need for greater forces in Tennessee against Union General William Tecumseh Sherman's army and ongoing campaigns against strong Union forces in Savannah and Charleston.

Efforts were made to fortify the area around Baldwin. The depot was a critical junction for rail traffic north-south and east-west to the Gulf at Cedar Key. East of Baldwin, Confederates developed a formidable fortress with several mile-long breastworks named Camp Milton on the west side of McGirt's Creek to blunt an attack coming west from Union forces in Jacksonville.

The Confederate leaders speculated about what it would take to dislodge the Union's fourth occupation of Jacksonville and the candid prospects viewed it as demanding a ghastly cost if it could be done at all. Their forces would need to cross significant bodies of water to get to Jacksonville, and nearby waterways were controlled by heavily armed Union gunships which would pound an approaching force. As containment became the settled strategy for the Confederates, there was another idea to deal with the free-roaming Union gunships.

What Confederates called "torpedoes" were deployed along the St. Johns and St. Mary's Rivers. Despite the name, these were more like explosive mines weighted to the river bottom. They would soon take their toll, sinking several prized river vessels like the large Union transport *Maple Leaf*. The Union Navy would eventually develop a scoop for the front of its ships to discover any of these "torpedoes" before they caused damage to a vessel.

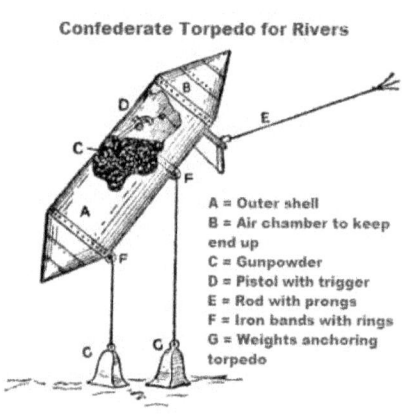

Confederate Torpedo for Rivers

A = Outer shell
B = Air chamber to keep end up
C = Gunpowder
D = Pistol with trigger
E = Rod with prongs
F = Iron bands with rings
G = Weights anchoring torpedo

A telegraph network was also developed, linking key Confederate locations across north and central Florida all the way to Tallahassee, including Waldo and Gainesville.

At this point, you should be introduced to a key Confederate figure in the story of the final year of Civil War action in Florida, and someone who plays a pivotal role in the story of the Marshall Plantation raid.

Chapter Five

Confederate Captain J. J. Dickison

Before the secession of Florida and the onset of the war, John Jackson Dickison was a resident of Orange Springs in northern Marion County where he had a highly valued plantation with eight slaves. He had migrated like many other South Carolinians to Marion County, Florida. The county itself was named after the South Carolina hero of the American Revolution, Francis Marion, known as "The Swamp Fox of the Revolution." Indeed, biographers would later call Dickison "The Swamp Fox of Florida" since he would replicate the same guerilla tactics in his role as a cavalry commander.

Capt. J.J. Dickison

With the secession of Florida and the outbreak of war, Dickison quickly offered his services to the Confederacy and would begin with an artillery unit in Virginia as First Lieutenant. Six months later, after failing to secure a preferred leadership position, he organized Company H of the 2nd Florida Cavalry known as the Leo Dragoons, becoming its Captain. The unit had the responsibility for protecting the north central Florida area from occasional Union incursions and raids. These became more frequent and troublesome with the fourth occupation of Jacksonville in February 1864.

Within days after the Union forces secured Jacksonville in February 1864, the 40th Massachusetts Mounted Infantry (MMI), a White unit, made far-ranging incursions picketing into the interior west of Jacksonville to test the strength of any Confederate units. A select unit of about 50 men drawn from three companies in the 40th MMI was dispatched to Gainesville. On February 14, they surprised over 100 of the town's ad hoc militia defenders who dispersed upon their arrival. The Gainesville guard may have heard of Union troops having landed in Jacksonville, but could hardly have expected them to arrive in Gainesville over 60 miles away.

The Union force under Captain George E. Marshall (no known relation to the Marshall family that owned the raided plantation) began assessing their situation as they secured Gainesville.

They found themselves quickly inundated by dozens of slaves seeking sanctuary and freedom. One of them coming into town warned the Union captain of a Confederate cavalry force of a hundred or more gathering to launch a counterattack.

Captain Marshall immediately put everyone to work, soldiers and escaping slaves, erecting hasty barricades from cotton bales stored in a nearby warehouse to create defensive positions. They only had about 40 minutes before the Confederate cavalry unit, commanded by Captain J.J. Dickison, began their assault on horseback.

Likely believing their superior numbers would overwhelm the small Union force, the Confederate cavalry charged the flimsy defensive line. What Dickison did not know was that the 40th Massachusetts Mounted Infantry had been outfitted with a new weapon that Dickison had likely never encountered before.

The Spencer 7 carbine has its own story. Invented by Christopher Spencer in 1860, its value as a military weapon was obvious as the war started. The Spencer 7 had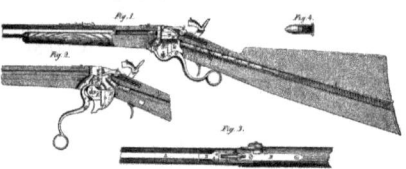

Spencer 7 diagram

a 7-round magazine operated by lever action to expel a spent round when depressed and chamber a new round from the magazine when returned. The magazine was a tube in the rifle's stock. The hammer needed to be manually cocked before pulling the trigger to fire each newly chambered round.

The Spencer 7 could produce an exponential increase in the rate of fire over the standard issue musket. A skilled musket operator could manage just 3 rounds per minute. The Spencer 7 could fire at 10-14 rounds per minute. (Later models had a replaceable magazine that could increase the rate of fire to 20 rounds per minute.)

This revolutionary weapon was certainly impressive, yet the War Department refused to procure them. Spencer tried repeatedly, working his way through every level of command, and got nowhere. He finally got a chance to demonstrate the weapon to President Lincoln himself on the White House lawn. Lincoln was thrilled and ordered that it be issued to every soldier.

Nonetheless, the War Department still refused the Commander-in-Chief's directive. The procurement bureaucracy cited the rapid rate of fire and believed that ammunition would be spent recklessly, making it impossible to keep ammunition supplied to soldiers.

Supply and logistics were always a challenge for large armies like those fielded by both sides in the Civil War, however, one senses that the bureaucratic wall was simply unyielding when such a powerful weapon was available that offered an outsized potential to bring the war to a conclusion more quickly.

The War Department did agree to begin outfitting cavalry units with the Spencer 7 due to the rifle's smaller size, portability, and suitability for use by a horse rider. On horseback, a cavalryman could not manage a musket, and single-shot carbines like the Sharps and Burnside models were still cumbersome to reload. The Spencer 7 was ideal for cavalry use.

The 40th Massachusetts Mounted Infantry was armed with Spencer 7s and Dickison surely had no idea what was in store for his charging assault at Gainesville.

A reporter from the *New York Tribune* embedded with the unit recounts the action in an article that would later be re-printed and published in the *Chelsea* (MA) *Telegraph and Pioneer* on March 5, 1864.

The first wave of Confederate cavalrymen was hit hard by Union fire as they came upon the barricades, and those whose horses jumped the barricades must have been horrified to find a full enfilade of Union fire immediately greeting them on the other side from the same rifles. The attackers were decimated as over 40 men were killed in a minute's time. The remainder of Dickison's cavalry withdrew from the brief catastrophic firefight, yielding Gainesville to the small unit of very well-armed Union soldiers with their Spencer 7s. There were no Union casualties, although two Union pickets had been captured prior to the attack.

A huge storehouse of material was found by the Gainesville train station, awaiting shipment north to Confederate forces. Unable to seize the vast stockpile themselves, Capt. Marshall let the heavily deprived townspeople take what they wanted. After holding Gainesville for 56 hours, the select unit departed with 36 freed slaves. They may have then rendezvoused with their main unit to join in the Battle of Olustee on February 20.

Dickison would return to Gainesville in August 1864 to deliver Union forces a stunning and stinging defeat, routing a superior force and taking hundreds of prisoners. However, in Dickison's volume of Florida actions in the Civil War, he devotes his attention in February 1864 to the unfolding of the Olustee battle and what happened afterward, never mentioning the awful encounter of his own unit in Gainesville on February 14 prior to Olustee.

As 1864 unfolded after Olustee, with a strong Union presence remaining in Jacksonville to threaten the interior, and also utilizing the St. Johns River for its deep incursions, Dickison realized that his small unit could have an exceptional impact by utilizing guerilla tactics. Using the ready support of locals to provide information on Union troop movements inland and ship movements along the St. Johns River, combined with their own knowledge of the terrain, Dickison and his unit created opportunities for favorable engagement, even against vastly superior Union forces. Having knowledge of the Union movements, Dickison could pick his preferred location for engagement, typically resulting in a sudden, bewildering ambush for Union units.

Dickison was soon well-known to Union troops who nicknamed him "Dixie" and "the Grey Fox." Indeed, the west side of the St. Johns River was known among Union soldiers as "Dixieland," the territory where Dickison was most dangerous to Union operations. (Dickison proved highly problematic to Union operations on the east side of the river as well.)

While there are many examples of his strategic guile, the best is also Dickison's most famous exploit. It is (unsurprisingly) the only time that a cavalry unit sunk a US Navy warship. Yes, Dickison's cavalry unit sunk the *USS Columbine* on the St. Johns River in what is called the Battle of Horse Landing.

Union forces occupied the nearly deserted riverfront town Palatka in April 1864, staying for six weeks. They began conducting raids along the St. Johns River as far south as Lake George. (Later in July 1864, Union forces would return to occupy Palatka deploying men of the

3rd US Colored Troops and 34th US Colored Troops from which the majority of the raiders were drawn.)

During this April 1864 period of occupation, Union officers found good hospitality from the Sanchez sisters of Palatka.

The Sanchez family was originally from Cuba, became wealthy in Florida, and did not appear to have had much interest or involvement with the war, although a son served in the Confederate army.

However, in April 1864, Union soldiers arrested patriarch Mauricio Sanchez on (false?) charges of espionage and had him imprisoned at St. Augustine's historic fort Castillo de San Marcos, renamed Fort Marion. It seems the Sanchez sisters now had a side to take.

Despite the arrest of Mauricio, Union officers still gathered on May 23, 1864 for the fine and friendly hospitality of the Sanchez sisters who still had to make ends meet.

On this occasion, these officers let slip that there was a planned ambush of Confederate forces (more precisely, Dickison's forces) amid the darkness of the next morning using the gunship *USS Columbine* which was moored off Welaka, a small settlement north of Lake George at roughly the confluence of the Ocklawaha and St. Johns Rivers.

The 117-foot sidewheel steam gunship, a refitted tugboat, with its two 20-pound rifled Parrott cannons, would be transporting over 100 Union soldiers of the 17th Connecticut Infantry as well as soldiers of the 35th USCT (former 1st NCCV) in addition to its usual 25-man crew which included Black seamen. The *Columbine* would be leaving Welaka and heading north that night to disembark its soldiers at Horse Landing, south of Palatka. These Union soldiers would then seek out Dickison's slumbering camp and ambush them before dawn.

Hearing this insight into the Union plans, Lola Sanchez had her sisters make excuses for her absence while she rode into the night to Dickison's lines to inform him of the Union plans. It wasn't far and, after delivering this key intelligence, she returned apparently without any notice taken of her absence about 90 minutes later.

Bear in mind that just the previous evening, May 22, 1864, Dickison's unit with its two 12-pounder field cannons attacked the heavily armed *USS Ottawa* from the dark shore, inflicting significant damage. Dickison's forces beat a hasty retreat when the *Ottawa's* five heavy cannons began returning fire, targeting the Confederate muzzle flashes on the shore. Still, the fine targeting of the cannon fire by Dickison's artilleryman Lt. Bruton damaged the *Ottawa* and caused it to languish in place for 30 hours while repairs were made.

Then Dickison raced south to reposition his cannons near Palatka. He arrived minutes too late to take a few shots at the *Columbine* as it sailed further south to Welaka. His second chance was coming.

It was that *next* evening, after surprising and hammering the *USS Ottawa* at Brown's Landing the night before, that Lola Sanchez gave Dickison the tip about the *USS Columbine's* early morning raid plans.

With this valuable intelligence, Dickison moved quickly to Horse Landing in the hours after midnight, concealing his sharpshooters and cannons at the landing amid the wild growth along the riverbank. And they waited.

Attack on the USS Columbine

At about 3 am according to Dickison's account, the gunship arrived as Dickison's disciplined men awaited along the dark shoreline for the command to open fire. The *Columbine* came within 60-100 yards of the landing when Dickison had his cannons and riflemen begin firing. Almost immediately, according to the Ensign in command, the *Columbine* lost its steering due to cannon fire while its decks loaded with soldiers ready to disembark were raked with rifle fire from the shore. The gunship drifted 200 yards away before coming to rest on a mud bank in shallow water. After forty-five minutes of fighting, the crippled *Columbine* surrendered.

Dickison rather dramatically presented the numbers, reporting that, of the 148 on board the gunship, only 66 survived and one-third of them were wounded. Dickison's math suggests heavy Union losses, however, Ensign Frank Sanborn noted 17 killed, only one being from the ship's crew. Many went overboard, several dozen drowned according to Dickison, while others swam to the east side of the river, eventually making it to safety in St. Augustine days later, and many others were captured.

Here is the report of Ensign Frank Sanborn commanding the *Columbine*:

> *I could discover nothing suspicious until directly abreast the landing,"* Sanborn said in his official report, *"distant about 100 yards, when two pieces of artillery, concealed by the shrubbery and undergrowth, almost simultaneously opened fire upon me. I instantly gave orders to 'hook on,' but unfortunately the second shot of the enemy cut my wheel chains, and at the same time the pilot abandoned the wheel and jumped over the bow. The vessel almost immediately went ashore upon a mud bank.*

After spending the day removing whatever was portable and valuable from the *Columbine*, including its dead, Dickison had the vessel set ablaze to prevent it from being recovered by one of the other Union gunships in the vicinity. Among the items removed from the *Columbine* were the orders from a Union general describing the plan to ambush Dickison's camp, a warning about sharpshooters, and instructions to prevent Dickison "at all costs" from crossing the river. Dickison had crossed the river to conduct several highly successful attacks on Union positions around Welaka.

Needless to say, Dickison had once again turned the tables on the Union forces, using an informant's espionage and knowledge of the landscape to set his own ambush.

In August 1864, Dickison and his men had been following a Union raiding party from the 4th Massachusetts Cavalry which had moved ahead of a 5,000-man main body toward Gainesville. Bypassing the stationary main body, Dickison with under 200 men surprised the Union occupiers in Gainesville. Estimated at roughly 340 men, these Union soldiers may have been dispersed around the town engaged in acts of looting rather than establishing secure lines, leaving them unprepared for Dickison's assault.

After two hours of fighting, Dickison's attack sent them fleeing in disarray. Several dozen were killed and wounded and nearly 200 Union soldiers were captured while several dozen more limped back to the safety of Union positions. Union commander Colonel Harris believed he had been attacked by 600-800 Confederates.

Clearly, Dickison's small but potent force was the scourge of Union operations and had become an obsession of Union leadership. No one could catch him, and no one knew when he would suddenly appear and ruin an operation.

In early February 1865, only a month before the Marshall Plantation raid, there was another noteworthy series of incidents.

A Union headquarters report detailing intelligence about a large shipment of cotton from Braddock's Farm (near today's Crescent City) in Volusia County was received by the St. Augustine garrison's Union commander Lt. Col. A. H. Wilcoxson. The orders instructed Wilcoxson to send a raiding team to seize the cotton. Deciding to lead the operation himself, Wilcoxson took other officers and about 40 men (or about 75 men or 100 men, accounts differ) of the 17th

Connecticut Infantry (yes, of the same unit that was unfortunately aboard the *Columbine*).

Dickison learns about the raid at the farm near Dunn's Lake, crosses the St. Johns River with 50 men (or 80, accounts differ), and arrives in time to see ten large six-mule-wagons filled with tons of cotton and other seized goods heading back toward St. Augustine under Wilcoxson's command. Dickison chose the most advantageous place in the road and set his men to make the ambush. The surprise was complete as the Union raiders were mostly captured, a few were killed, a few escaped, and the wagons were seized.

Wilcoxson is dramatically wounded, having fired his pistol empty, thrown it, and drawn his sword, trying to attack Dickison who simply sought his opponent's surrender. With both on horseback, Wilcoxson charges Dickison with his sword. Like a Monty Python sketch, Dickison draws one of his revolvers and shoots Wilcoxson who then turns and attacks with his sword twice more, getting shot twice more before dropping from his horse. Wilcoxson would succumb to his wounds as a prisoner in Tallahassee a month later.

It would then take 25 hours for Dickison to complete crossing the St. Johns River to the relative safety of the west side with all the wagons, mules, and prisoners obtained in this mission plus his own soldiers and horses, in addition to several hundred men in another unit that arrived from their separate attacks on Union forces, using the lone flatboat available to him.

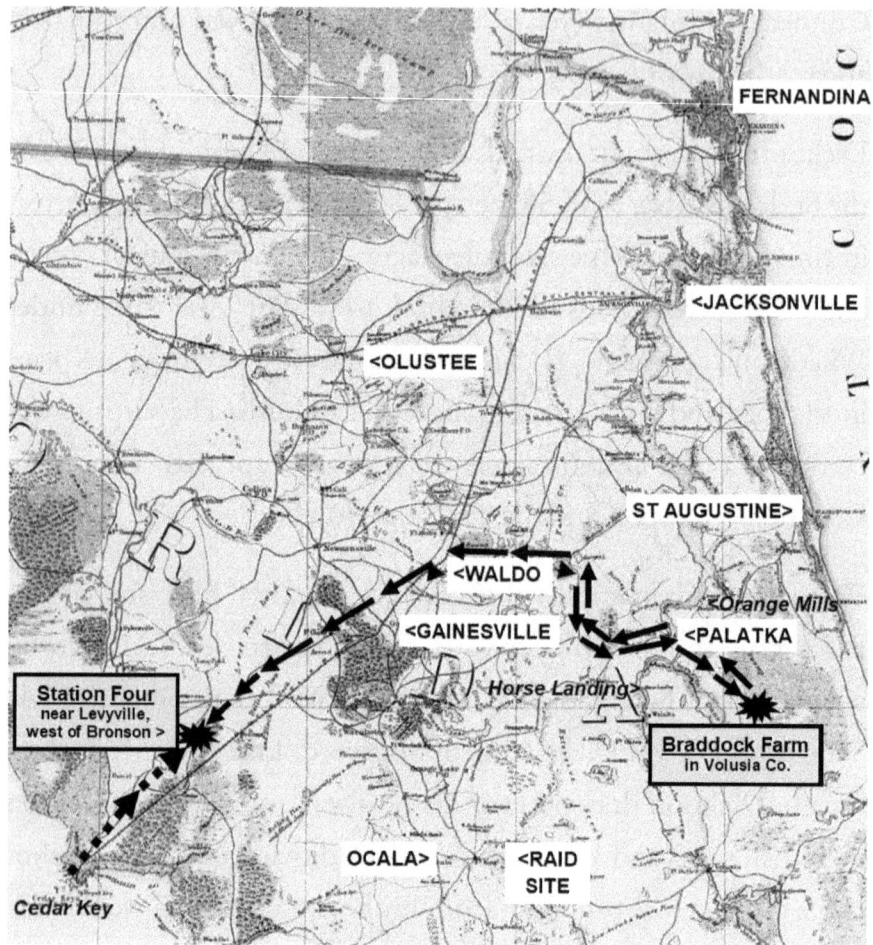

Dickison's movements in February, 1865

Yet when Dickison finally returned to Waldo, he was greeted with the news that a contingent of Union forces – about 400 men, with half being Black soldiers (former Florida slaves) – had landed at Cedar Key across the state on the Gulf coast. That was followed by a report that this contingent had penetrated to Levyville, west of Bronson in Levy County. Dickison gathered about 90 of his exhausted men plus the two cannons on February 13, 1865 and raced 60 miles across the state to intercept them.

Dickison, supplemented by local Home Guardsmen, blocked the Union advance, but the Union lines formed strongly. Four hours of fighting to a stalemate meant the Confederates had fired their last cannon round and were down to no more than three rounds of rifle ammunition per soldier. While Dickison's unit had to withdraw, the Union forces decided it was too difficult to pursue their mission and they also withdrew, returning to Cedar Key and departing on their ships.

Incredibly, Dickison had intercepted Wilcoxson's raiders on the eastern side of the state and then immediately raced to the western side of the state to intercept another substantial Union incursion. Covering actions on either side of the state in the same week is certainly noteworthy.

In his account, Dickison waxes majestically about this fight in Levy County, reflecting the "Lost Cause" theme in his writing in the 1890s. Remember that many of the Union troops were Black soldiers:

> *The enemy had advanced some distance in the interior, plundering the unprotected citizens, and were so insulting and brutal in their threats that the bravest hearts among our fair women trembled and lips grew pale at their approach. Had it not been for the timely arrival of our heroic little band and the brave militia soldiery who so bravely hastened to their assistance, fearful indeed would the result have been. Thank God who giveth the victory, "the battle was not to the strong," and the horrors that*

had again threatened every home were averted by His overwhelming love.

Chapter Six

Why Raid the Marshall Plantation?

There are a lot of questions to be raised about this wildly daring mission 100 miles behind enemy lines. Looking at the context, it seems at first glance like pure madness.

Let us recognize that the Civil War effectively ends for the Confederacy with the loss of its primary fighting force following Confederate General Robert E. Lee's surrender to Union General Ulysses S. Grant on April 9, 1865, at Appomattox Courthouse in Virginia. Fighting will continue in various areas, and the Confederacy will hold out a few weeks longer, but the main Confederate army and meaningful opposition to Union forces are finished. The clock is simply ticking down on the Confederacy after Appomattox.

The inevitability and imminence of the end of the war is fully understood in March 1865. Yet the Marshall Plantation raiders left Jacksonville, Florida on March 7, 1865 to embark on their extraordinarily risky mission. They were surely aware of the status of the war.

In mid-February 1865, young – only 26 – Colonel George Armstrong Custer in the command of Union General Philip Sheridan's cavalry corps defeated the demoralized and impoverished Confederate forces of General Jubal Early, effectively driving the Confederate army out of Virginia's Shenandoah Valley.

Lee was now going to be pressed by Sheridan from the west. From the north and east, Grant's armies advanced, having laid siege to Richmond and Petersburg. Meanwhile, to the south, Union General William Tecumseh Sherman was making stunning strides in his march northward through North Carolina as he closed in on Virginia's southern border. Lee had nowhere to go and no help on the horizon.

Everyone surely knew the war's end was just weeks away. Yet the raiders set off on their seemingly reckless mission.

Why? Did the raid have strategic importance?

The Union army's Department of the South was at best of tertiary concern to the Union strategy in prosecuting the war, but it is fair to state that Florida was largely out of any strategic military picture, Union or Confederate.

Early on, with the first Union occupation of Jacksonville in 1862, Confederate General Robert E. Lee, then the commanding officer for the southern states prior to commanding the Army of Virginia, told Floridians that they were on their own for their defense; the Confederate military simply would not be responding to Union forces in Florida.

The early Confederate apathy toward Florida would change by the fourth occupation of Jacksonville in 1864 as we have already reviewed in the Confederate scramble to get reinforcements to Florida in time to meet a large Union force at Olustee. Supply and logistics had radically changed for the Confederacy and a response to the new Union threat in Florida demanded a substantial force.

Union force levels (and interest) in Florida surged and ebbed, sometimes without reason, like General Hunter's abrupt and inexplicable withdrawal of Union forces from Jacksonville in 1862. Typically, drawdowns came from a demand for more soldiers for either the Virginia or the Sea Island campaigns. Florida was one of the lowest priorities for the Union military.

Florida served as a key supplier to Confederate forces, becoming more critical when Grant seized Vicksburg in July 1863 which severely limited the Mississippi River for Confederate supply lines to its units in the west. Florida's importance to the Confederate supply chain grew dramatically after Vicksburg's fall.

Union missions in Florida mostly consisted of raiding and trying to disrupt the Confederate supply lines and logistics, while freeing slaves which deprived plantations of their labor and provided potential Union military recruits. Using the Union forts dotted around the coast as well as inland incursions, Union operations often utilized the St. Johns River and the largely unchallenged Union Navy. While Tallahassee was much coveted by Union leaders, it was rarely even threatened.

Whatever happened in Florida would have a scant impact on the outcome of the war in any case. This odd Florida raid of a distant Marion County sugar plantation was inconsequential strategically.

Fitting within this context, the target location of Marion County was so far south and inland that it was not a real factor in any military strategy at any time during the war, at best a distant thought. While Marion County was a major Florida supplier of agricultural products for the Confederate army, it was hardly considered a target because it was so far off the strategic map. The closest that the Union forces came to Marion County was a couple of Union operations in Alachua County to its north, like around Gainesville, and the raids on the St. Johns River which came as far south as Lake George along Marion County's sparsely populated eastern border.

Yet the raiders seem to have chosen Marion County and the Marshall Plantation intentionally. The raiders certainly didn't end up there by accident. So, why target Marion County and the Marshall Plantation? And why in March 1865?

These questions have taunted the curious, and answers would surely provide insights to explain what these raiders were trying to accomplish.

There would be one more significant action in Florida in 1865 before the war's end.

Union forces made one last attempt to seize their long-sought prize, state capital Tallahassee, in early March 1865 – the only Confederate state capital not seized by Union forces. Coming up the narrow, shallow St. Marks River south of Tallahassee, Union forces disembarked and met Confederate opposition at Natural Bridge.

Facing a depleted, ad hoc force comprised of Confederate soldiers, old men, young boys, and cadets from Florida Military and Collegiate Institute (later to become Florida State University), the superior Union forces on March 6, 1865 were once again stalemated by the determined defenders entrenched in excellent positions. Fighting all day, the Union forces could not forge a way forward. They returned to their ships, and Tallahassee was left untouched yet again.

While this Union operation failed at Natural Bridge, occurring also in March 1865, we understand the symbolic importance of seizing Tallahassee, the state capital, even though there is no real strategic point at this stage of the war. The Marshall Plantation raid had no symbolic importance.

The Marshall Plantation raiders of our story will depart Jacksonville on March 7, 1865, just a day after the Union withdrawal from the Battle at Natural Bridge, as this Marion County raid marks what seems to have been one of the last or was the last planned Union operation in Florida in the Civil War.

Setting the context has shown that Union operations in Florida were largely ineffectual. Despite superior forces and resources, the defenders of the Confederacy enjoyed strong loyalties among the people in the interior which, coupled with the advantage of operating defensively on their home turf, enabled their smaller and less equipped forces to manhandle many Union efforts.

While Union raiding activities could be successful, they required significant deployments and substantial support to succeed. Even then, units were often harassed and beset by Dickison's aggressive and daring cavalry.

Union forces never extended their geographic control of Confederate territory in Florida beyond the confines of their few forts and garrison towns except for temporary occupations.

A great deal was shared earlier about Capt. J. J. Dickison because of his renown among Floridians and among both Confederate and Union troops for his clever use of the resources available. His wily successes and ability to escape capture were as annoying for the Union leadership as was their inability to capture Tallahassee. No one seemed to be able to best "Dixie" – Dickison's nickname among Union soldiers – any more than they could capture Tallahassee. While the territory on the east side of the St. Johns River posed less threat to Union operations, Union soldiers nicknamed the lands on the west side of the St. Johns River "Dixieland," a nod to the Confederate captain whose threat was pervasive on the west side of the river.

To sum up, Union operations were consistently disappointing while Dickison's exploits were consistently celebrated by his allies as well as reviled by his Union opponents. Could Union forces in Florida end the war with both Tallahassee and Dickison unscathed? Would Black soldiers finally have their own role to play?

Chapter Seven

Critical Intelligence and a Unique Strategy

Sergeant Harmon's letter discovered by reporter Rick Allen provided plenty of hitherto unknown details of the raid, but the letter does not explicitly state its mission strategy. Piecing together the details can give us the basic outlines of the operation which can be tied to Harmon's comments for confirmation.

Typical Union operations utilized waterways, particularly the long, large St. Johns River, to send gunships guarding transports to deploy and extract soldiers in raiding the interior. There was no secret about Union movements; Union leaders knew that. Union raids had proven sufficiently successful and disruptive, even though any territorial gains were temporary at best. Largely, Union forces were contained around Fernandina, Jacksonville, and St. Augustine in northeast Florida in January 1865.

The terms of engagement of the Marshall raid were exceptional. The raiders would secretly move south on the St. Johns River, traveling 100 miles from Jacksonville to conduct their raid. There were no gunships to guard their passage and no gunships to enable their extraction. They would be on their own going to the target and on their own returning to safety. That meant a return trip traveling on foot to the Union garrison at St. Augustine 80 miles away from the raid target.

On its face, it seems insane – a good way to get killed or captured. And why do so when the war is nearly over?

The raiders developed a strategy that re-imagined the St. Johns River. Where the White military leadership had counted on gunships and seen the river as a transportation asset to be exploited, the raiders could envision it differently.

A Navy gunship would be readily observable moving along the river with smoke billowing skyward from its engine. However, the river was also wide, largely unpopulated, and offered plenty of hiding places along its wild banks. This could make the movement of a small unit in small boats difficult to discern, and if done well, impossible to detect. The element of surprise could be leveraged in an operation using this strategy.

The raiders also realized that the river was a mile-wide wall of water. This wall of water was a huge issue for a cavalry unit like Dickison's which required a flatboat to move its horses. A flatboat was also required for crossing wagons and any large animals.

There were few places for a cavalry unit to make a river crossing along the largely wild, swampy, and unpopulated St. Johns River. A cavalry unit needed established landings on both sides of the river to get across successfully. There were three of interest on the southern half of the river: Palatka, Horse Landing, and southernmost at Fort Gates where a ferry operation had been ongoing since 1853.

It was likely common knowledge that Dickison had at least one flatboat available to him. However, finding that flatboat along the wild, dense shore of the St. Johns River was highly unlikely. It was well-hidden, and it was so valuable that it may have been guarded as well. It would have been positioned close to Palatka for fairly rapid access from their base in Waldo.

Flatboat carrying a wagon on the St. Johns River

The second crossing location was Horse Landing, the site of the demise of the *USS Columbine* at the hands of Capt. Dickison related earlier. Its sunken hulk remained off the landing.

The third crossing location at Fort Gates was just south of Welaka, a village on the eastern side of the St. Johns River, well south of Palatka and not far from Lake George.

Welaka had suffered during the later years of the war. Its population was 180 in the 1860 Census, but only 20 remained by 1865. Welaka had been the scene of a brief Union occupation beset by maddeningly difficult encounters with Dickison's unit. His Confederate cavalry crossed the river to harass the Union forces on the river's east side several times. This provoked the Union's abortive attempt to ambush Dickison's sleeping Confederates with the soldiers borne on the Columbine. We know how that worked out.

One can easily imagine that the Fort Gates ferry was not commonly operational with Union gunships on the waterway and raids being conducted nearby. It is quite likely that a flatboat was hidden for local travelers to utilize when Union forces were not present as a threat. In any case, there were established landings on both sides of the St. Johns River at Fort Gates.

With a team of "scouts" among the complement of raiders, the scouts had likely gained intelligence, possibly from slave sources on a nearby plantation, disclosing the location of the hidden flatboat/ferry. It could be seized for use by raiders when they needed it.

This crossing point at Fort Gates would situate the raiders in a desolate area on the west side of the river marked on maps of the time simply as "Extensive Scrub." Barely populated with a few wagon paths, any movement would naturally be quite uninhibited and undetected. They could expect to encounter little on their way to the bridge at the Ocklawaha River about 30 miles southwest of Fort Gates. Unseen meant being unreported to Dickison.

By the way, Florida's interior was still largely unexplored wilderness, and maps were few, poorly drawn, and wildly inaccurate, particularly Union maps. Again, scouts were invaluable to the Union military in getting an understanding of the terrain that an operation would be encountering.

The Marshall Plantation was located about one mile west of the Ocklawaha River bridge, a major sugar cane operation with dozens of slaves producing a variety of valued supplies. It was a decent target for this remote area.

However, another piece of intelligence may have been secured by the scouts that put the bulls-eye on Marshall's. The scouts may have learned that, due to plantation labor needs, there was a large influx of additional slaves to carry out the work – dozens more – possibly brought in from the Marshall's cotton plantation in northwest Marion County and possibly further supplemented with leased slaves. This would make the Marshall Plantation a very tempting target.

Of course, the attentive observer is still asking why they chose this target so far away. It is so far south and inland, much further than any Union force had ventured, indeed quite beyond any previous consideration.

Yes, there is a reason.

Having determined that Dickison has one primary crossing point at Palatka, the raiders may have then calculated that, if he could be drawn from his base camp around Waldo sufficiently far south of Palatka in pursuit of the raiders, say to Ocala or Silver Springs, or to the Marshall Plantation itself, or even into the scrub on the east side of the Ocklawaha River, the raiders could then make their escape in the opposite direction. By the time Dickison arrives in Marion County and receives any update, the raiders would be across the river and headed toward St. Augustine with a big lead over their likely pursuers. Dickison would then have to make the trek from Marion County back to Palatka to cross the river to begin a pursuit of the raiders on the east side of the river. By the time Dickison could do that, the raiders expect to be well on their way to St. Augustine.

The strategy of drawing Dickison out of his preferred position was very risky. Telegraph communication between Ocala where the Home Guard was based and Dickison's base camp in Waldo was assured. How quickly communication about the raiders reached Dickison and what information would be provided was a huge question mark. If Dickison delays leaving Waldo, or word somehow gets to him about the raiders' direction of flight while Dickison is en route south, he could change course and head directly to Palatka, cross the river, and then intercept the fleeing raiders quite easily.

Of course, the raiders had to rely on their own stealth to avoid being detected as they traveled the St. Johns River from Jacksonville all the way to Fort Gates and then overland to the Marshall Plantation. They may have also relied on the initial communication about a raid in Marion County to shock Dickison into hasty action drawing him south.

Remember, Dickison had had the benefit of his information resources among the residents espying Union movements on land. On the river, his people would notify him of any gunboat expeditions. In the scenario of the raiders' strategy, he would be getting his first knowledge after the raid had happened, learning that Union raiders had successfully penetrated exceptionally deep behind his lines without anyone knowing about it and telling him. That had never happened before; indeed, no Union raiders had ever ventured to Marion County.

For Union forces to appear that far south, standard Union operations would have dictated a gunship, easily observable on the river by Dickison's lookouts. In the raiders' scenario, there would be no gunship, no warning, and hopefully, no ability for Dickison to write the script for ambushing the Union force.

Panicked by word of such a raid, the raiders seem to count on Dickison leaving Waldo in haste. He was unlikely to wait for an updated report when he knew there were raiders freely roaming and operating deep in his backyard, wreaking who knows what kind of havoc and damage. He would have to act fast. Only his unit could provide a sufficient counter-force.

There is a slim possibility that an updated report could reach him as he passes through the countryside, let's say in Gainesville, but it is a slim possibility. There is no likely way that he could be expected to get new information until he reaches Ocala-Silver Springs. If so, then Dickison would be just where the raiders want him to be; that is, far from Palatka, still on the west side of the river.

Meanwhile, the raiders, having crossed the river at Fort Gates to the east side of the river, would be moving on their way to St. Augustine.

The raiders' plan could work.

More on the Reason for the Raid

We have unpacked the raiders' strategy, but questions remain, and points need to be identified.

The plan for the Marshall Plantation raid originates with the raiders, not with the Union command. As noted earlier, this seems to have been one of the last organized Union military actions in Florida.

General U.S. Grant had pulled forces from the Department of the South in August 1864 to supply the Army of the Potomac, as had the newly appointed X Corps commander, Major General Gillmore, removing thousands of soldiers from Jacksonville, impinging on operations to the point that nothing was really planned in Florida. General Halleck's 1865 report from the Department of the South indicated that Florida operations were "purely on the defensive." The war is virtually over, and no significant operations are planned for the irrelevant Florida theater. All eyes focus on Virginia. In the Union's Dept. of the South, the focus was on taking Charleston SC and Savannah GA as General Sherman marched to the sea and then headed north into North Carolina. Well, except for those who wanted to try one more time to take Tallahassee ... and fail.

Further, the plan for this raid has none of the hallmarks of Union strategy. The risk of sending a unit deep into enemy territory on the west side of the St. Johns River without a heavily armed means of extraction would hardly be considered. Indeed, it would be dismissed by the Union command as consigning such a unit to be killed and captured. Such a plan was not worth considering at any time by the Union command, and certainly not with the war nearing its end.

Interestingly, the nature of the Union command is one of the key drivers for the raid. For years, Black soldiers like the raiders had received their orders from their White officers and carried them out. There was likely little input from the non-commissioned Black officers – sergeants – that was either sought or desired by their White commissioned command officers – captains, majors, and above – about the conduct of operations. The raiders knew that they were quite capable of organizing and leading their own operations, but they never had the opportunity to do so. White commissioned officers led all operations, reflecting the entrenched belief that Black soldiers were inadequate for the leadership task simply because of their race.

Having developed their own unique plan, the raiders wanted to lead the operation as well. One of the key points cited by Sgt. Harmon near the end of his letter said that this operation was led and carried out by Black leaders and soldiers. Here is the revealing sentence that stands as Harmon's summary comment after having described the details of the raid:

> *This expedition reflects great credit on Sergt. Major James, for the masterly manner in which it was commanded, and gives further proof, that a colored man with proper training can command among his fellows and succeed where others have failed.*

That is a justifiable source of pride. The raiders wanted to prove their mettle to their White superior officers, that Black leaders and soldiers were just as capable of organizing and executing operations as their White commanders. They wanted respect and the acknowledgment of their abilities from their White officers which had been lacking throughout their term of military service.

This was intrinsic to the difficult journey of Black soldiers since they grew to become a critical part of the war effort in mid-1863. At first, they were disrespected with "fatigue" duties. Then they proved themselves on the battlefield, demonstrating time after time in combat that they were equal to any White soldier. Whether it was dashing through the curtain of vicious gunfire on the open beach to attack Fort Wagner, or being green and inexperienced as your unit gets decimated at Olustee but still standing tough for hours of fighting and learning how to fight in the hardest way imaginable, Black soldiers had proven their combat worthiness to their White counterparts and commanding officers. The hearts and minds of the White military, regardless of their personal opinions about Black soldiers before they were engaged in combat situations, would come to have no complaint about the courageous and determined fighting capability of Black soldiers, only compliments and admiration.

Now the war was almost over. Mission planning and operational leadership still remained beyond the reach of Black soldiers. The men of the 3rd USCT and the 34th USCT had seen plenty of action, but mission planning and operational leadership – the leading functions – had been denied them up to this point. If these veteran soldiers were going to prove themselves in leadership, in their planning ability, in their self-operating capacity apart from White commanders, and show what they were truly capable of, time was running out.

One final point is critical to identify. It is presented in Sgt. Harmon's letter with this final clause, the title of this book: "… and succeed where others have failed." To which "others" might he be referring, and to what "failure"? In the context of the raid being led by a Black Sergeant-Major, the phrase cannot refer to Black leadership having ever failed; they had never had the opportunity before. This 'failure' can only refer to White command leadership.

This phrase seems a direct reference to the Union command's inability to pull one over on Dickison. Dickison consistently beat the Union command with a force that was typically much smaller than the Union's deployment. Failure was writ large over the Union's efforts to overcome Dickison since Olustee. That failure was owned by the White Union command, but Sgt. Harmon would never think of stating that fact so directly. When he wrote the letter, he was still in the US Army after all. Great unpleasantness could result from indelicate wording.

Perhaps the most salient point for the raiders is to beat Dickison, to do it themselves, using their strategic plan, being led by their Sergeant-Major, and hope to succeed where others have failed. While the raid would aim to destroy a significant plantation, disrupt infrastructure, and liberate a large contingent of slaves - all of which was important - beating Dickison is also a primary focus of the mission. No one had outfoxed the sly and slippery Dickison. The Black Union raiders set out to show they could succeed where others have failed.

Chapter Eight

Getting Started

We have reviewed the raiders' strategy and mission, but we ought to clarify how it likely came about. With no source disclosing this information, we must use some logic.

The key factors were the availability of a ferry/flatboat for crossing the St. Johns River at Fort Gates, and the doubling of the slave population at Marshall's sugar plantation, intelligence that was certainly gleaned by the scouts who would become part of the raider team.

Assuming a successful journey unnoticed and without incident traveling the St. Johns River from Jacksonville to Fort Gates, then the success of the raid hinges mostly on moving from the west side of the St. Johns River where the raid gets conducted to the east side – opposite the expected pursuer Dickison and thereby setting the mile-wide river between them. That requires a flatboat if it involves anything of greater size than people, i.e. wagons, horses, and mules, likely booty from a plantation raid. The wagons, horses, and mules would aid in the escape back to safety.

A flatboat would be of considerable size, but hiding it along the wild shores of the St. Johns River would not be a problem.

How large would a flatboat be?

Dickison's flatboat, according to his accounts, had the "capacity to carry one wagon or twelve men and horses." Despite its size, Dickison's Palatka area flatboat was never discovered to our knowledge, remaining available to him throughout the war. Without knowing precisely where it was located, one would likely never find it.

We can assume that the flatboat at Fort Gates was of similar size to Dickison's, nothing exceptional, minimally able to handle a wagon and perhaps a team of horses.

An additional piece of intelligence garnered by the scouts provided the target for a mission. Likely receiving the intelligence about the flatboat from a slave on a nearby plantation, the scouts may have also learned that the slave population at the Marshall's sugar plantation was exceptionally large.

The 1860 slave census for Marion County noted the two plantations owned by Jehu Foster Marshall, one in northwest Marion at Wetumpka - a cotton plantation - and the sugar plantation near the Ocklawaha River in eastern Marion. Each plantation had 40-odd slaves listed respectively, the plantations' typical working slave population.

The scouts likely learned that the slave population at the sugar plantation had roughly doubled, likely due to seasonal labor needs for clearing, planting, or processing products. It seems likely that the slaves at the northwest Marion cotton plantation were brought to the sugar plantation in southeast Marion for this extra workload. The plantation managers may have also leased slaves from either a nearby plantation or from slaveowners who had become refugees in Marion County, having once had their operations located in contested territory in northeast Florida that were ruined by the war or abandoned in fear. These slaveowners would bring along their valuable slaves in their flight deeper into Florida's interior, affording the opportunity to earn income by leasing the slaves.

The flatboat location together with this information about the surge in the slave population made the sugar plantation a tempting target and explains the mission.

Knowing that a river crossing could be made at Fort Gates with a flatboat available nearby, and knowing that a plantation in the vicinity would have an exceptional number of slaves present, the next speculation would concern how this important knowledge could be utilized.

The scouts likely shared this knowledge of the flatboat and the increased slave population at the Marshall Plantation with White officers. It seems the White officers did not know what to do with it and/or did not think that, at this waning stage in the fast-closing war, it was actionable. However, when the raiders heard about this inside information, their minds went to work about how to exploit it.

When this key intelligence became known to the raid planners, and when the pieces of their plan were assembled, and when the whole strategy took shape seems difficult to determine. Planning takes time and we can be sure that the raiders took sufficient time to figure out how long things would take to happen and how likely events were to unfold, plus measure all the things that could go wrong. The scouts would also have valuable input to contribute besides the intelligence gathered. It would come together rapidly in any case. How long the slaves would remain at the plantation and how long the flatboat would remain where it was known to be were factors that demanded quick decision-making.

We should appreciate the major influence of the scouts in this planning. It was their intelligence that located the flatboat and their intelligence that identified the exceptional number of slaves at Marshall's. We can also surmise that the scouts were very familiar with the routes being chosen: the St. Johns River going to the target, and the overland route going to St. Augustine. The scouts had likely been involved in numerous operations to move escaping slaves to Union sanctuary locations, either overland to St. Augustine, or north on the St. Johns River to Union-occupied Jacksonville. These two routes would have been very familiar to the scouts, major elements in the mission's probability of success.

Having made their determinations, it was time for the raid planners to present their strategic plan to their commanders and convince them to approve it.

Getting Approval

Once the raiders' plan has been finalized, for the most part at least, it would need to be approved by a White commanding officer like Colonel Benjamin Chew Tilghman, commander of the 3rd USCT and likely the Jacksonville garrison commander, or Major Frederick Bardwell of the 3rd USCT.

There is no way that the raiders simply slipped out of Jacksonville in the middle of the night with three pontoon boats on a mission unapproved by the commanding officers. They would be shot for desertion!

Being a regimental officer of the 3rd USCT, Sergeant-Major James would have ready access to either the Major or the Colonel.

We would imagine this White officer wondering some of the same things we have. The officer might ask if the raiders were aware that the war was almost over, that Marion County was 100 miles away, that there was no plan of extraction besides walking from Marion County to St. Augustine, that Dickison's cavalry would surely hear about it and pursue them on horseback with a strong force of veteran soldiers, that they would find it nearly impossible to fight off that force, likely superior in number, while also trying to protect dozens of freed slaves, and lastly that this whole plan was so risky and unnecessary – crazy! - why even do this?

Of course, the raider leaders would assure their White superior officers that they were indeed aware of all these considerations. Nonetheless, they would highlight the plan's well-considered feasibility, indicating how the team had considered the merits and demerits of the strategy, likely with testimony from someone like Chief Scout Israel Hall, mentioned in Harmon's letter, concerning the availability of a flatboat at Fort Gates, and noting the increased number of slaves present at the Marshall Plantation as well as their familiarity as scouts with the routes along the river at night and northward to St. Augustine through the barren scrub.

However, a key factor likely to influence their White superior officer would be the opportunity to best Dickison, conducting a raid deep in Dickison's own backyard and getting away with it.

This slippery, dangerous opponent who had wreaked havoc over Union operations for years might finally receive his comeuppance. The loss of Lt. Col. Wilcoxson and his men at Braddock's Farm only a month before due to a Dickison ambush on the return trip to St. Augustine was surely fresh in everyone's mind, followed by Dickison's subsequent repulsing of a Union advance from Cedar Key across Levy County at Station Four. This opportunity to beat Dickison would be quite tantalizing for any Union officer.

Further, the raider team would consist of volunteers for the mission. Each member of the team chose to participate in the operation. No orders were needed to staff the raider team, permission simply needed to be granted to those willing.

Finally, the raid planners indicated that they wanted to lead the operation themselves. It was their plan, their risk, and the unit's accepted responsibility; they did not need or want a White officer to lead the operation. They wanted to demonstrate their own capabilities whatever the outcome may be.

One could imagine the sigh of relief from every White officer at not being expected to lead this hare-brained, dangerous operation, having so far survived this awful war that was rapidly coming to an end.

Ultimately persuaded and supportive, we could expect the officer to seek approval of the operation from the garrison commander or the commander of the 34th USCT since the raider team would be a select composite from two Union army units plus civilian scouts.

Obviously, the final decision was the approval to proceed. There were no other operations planned or even contemplated. No officer was even desired for this mission. It concerned a small group of highly motivated soldiers and scouts who volunteered for this mission. Finally, the war was nearly over. If this group of men wanted to take one last shot at Dickison, the White command said, 'Go for it.'

We should also acknowledge that Sergeant Major Henry James was placed in command of the operation. James must have been a proven leader, having consistently demonstrated his leadership qualities to his White officers. Of the hundreds of Black men enlisting in the newly formed 3rd USCT, he was appointed to the rank of Sergeant Major, the highest ranking position for a Black soldier, having such authority for the whole regiment. Choosing him for mission leadership also meant that he was respected by his non-com peers and his men who would accept his leadership and follow his orders faithfully. These White officers show their confidence in Sgt.-Major James by placing him in charge as well as approving the daring mission under his command.

Assembling the Raider Team

Sergeant Harmon's letter [1] supplies few names of the raiders, but he does tell his readers about the unusual composition of the raider team. We can imagine that they were all thoughtfully recruited, likely a team of varied qualities, interests, skills, and experience. Here is Sgt. Harmon's description:

> ... an account of an expedition, which left Jacksonville under the command of Sergeant-Major Henry James, 3d U.S.C.T., on the night of the 7th of March consisting of sixteen (16) of the 3d U.S.C.T., six (6) men of the 34th U.S.C.T., and seven colored citizens, and one (1) of the 107th O.V.I. Thirty (30) men in all.

---

1. In previous editions, letter writer Sergeant Harmon was regarded as part of the raider team. A reviewer challenged that assertion. It seems one would have to agree that upon closer examination of the letter, it seems very doubtful that Sgt. Harmon was part of the mission. Had the letter writer been a raid participant, it would be expected that this would be reflected in the letter. While the letter is quite detailed and gives plenty of evidence of accurately reporting most of what happened during the raid, there is no instance of the letter writer describing events in either the first person singular (I/me) or first person plural (we/us), or associating himself with any of the action that occurred. It would be remarkable for the letter writer to omit his own activity, particularly given the enthusiasm Sgt. Harmon has for the raid.

While Harmon's letter is not likely a first-person account, it does seem to be the careful transcription of events during the raid from a first-hand participant. We can theorize that Harmon, aware of the gravity of the mission, sought to communicate this exceptional operation and convinced a member of the raider team, possibly Sergeant Major James but surely a fellow soldier in the 3rd USCT, to provide his account of the raid to be recorded and shared. The letter may not be a first-hand account, but its detailed depiction of events indicates that it is a solid account from a first-hand participant.

Now let us consider each component of the group detailed in the letter excerpt above.

Sergeant-Major James and Sgt. Joel Benn (and his younger brother Private Jerome Benn) all began their service in Company B of the 3rd US Colored Troops mustered in Philadelphia. (Again, Sergeant Harmon, the letter writer, was also in Company B, 3rd USCT.) We know that Sgt. Benn was part of the raider team and can assume that the 14 other soldiers from the 3rd USCT were from Company B.

With just over half of the raiders coming from Company B of the 3rd USCT, these were soldiers who had the shared experience of combat and service, having covered some of the target terrains during their missions in Florida. The St. Johns River and its nearby areas had been places of operation for both the 3rd USCT and the 34th USCT. There would not be many things unfamiliar to them.

The 34th US Colored Troops was formed in February 1864 when the early forming 2nd South Carolina Volunteer Infantry was re-formed and renamed upon arrival in Jacksonville. Those comprising the

2nd SCVI/34th USCT were freed slaves, mostly from Georgia and Florida.

Tens of thousands of slaves were freed in the South either through Union raids and military actions or through escape networks like Harriet Tubman's 'Underground Railroad,' or both, like when Tubman teamed up in a Union military action on a raid up South Carolina's Combahee River which extracted roughly 800 slaves who had been alerted via Tubman's network.

Having a unit of six men from the 34th USCT join the raider team seems unusual: why not have all 3rd USCT soldiers?

These six from the 34th were surely veteran soldiers who had seen plenty of action both inside and outside of Florida. They had conducted operations with the 3rd USCT, like the siege operations against Fort Wagner on Morris Island, South Carolina in the fall of 1863, and their shared occupation of Palatka the previous summer of 1864. The 34th had returned to Jacksonville the month before the raid, having participated in battles around Charleston including the intense battle at Honey Hill.

However, in speculating about the motivations of these six soldiers for joining this raider team, there is a wide range of options.

One is that these were former slaves who had escaped their plantation masters, likely in Florida where over 1000 former slaves were recruited into the Union army. It is even possible they were from Marion County plantations.

Slaves made up roughly 75% of Marion County's population at the beginning of the war, likely 7500 slaves. To have six escapees from Marion County in the 34th requires no imagination.

Another possible motivation could have been their negative experience in encountering Dickison's cavalry during their Florida actions.

The key reason was probably their training and experience in the "jayhawker" style of engagement which can be seen in aspects of the mission. If the mission plan calls for burning a plantation to the ground, these were soldiers who knew how to get that job done. They had burned targets thoroughly, from plantations along the Combahee River in that raid involving Harriet Tubman to the inexplicable burning down of Darien, Georgia under orders from their commander, Col. Montgomery.

Whatever the motivation, six soldiers of the 34th USCT, freed slaves, quite possibly from Florida, maybe even from Marion County, veteran "jayhawkers," are part of the raider team.

Sergeant Harmon's letter notes "seven colored citizens" as part of the raider team. In another place describing the raid's action, he refers to "scouts" and then separately in the same sentence to a "citizen."

The confusion of twin labels is not surprising. A "citizen' is not a military person or under military orders. "Scout" per se seems to be an informal designation that straddles being a "citizen," but who also actively (and surely with compensation of some sort) assists the military. Given the reluctance of many freed slaves to serve in active Union frontline military roles, a way of supporting the Union's military effort without being a uniformed target under a White officer's control would be to serve as a civilian scout.

Information about Union "scouts" is extremely thin. They were not "spies" per se, pretending to be one thing for one side while actively reporting to the other side. Rather scouts were conducting surveillance and providing key intelligence to the Union military. They used their access to sources and inside information plus their invaluable familiarity with an area to provide counsel to military leaders.

For Black scouts [2], we would expect that they had contacts and informants working in various occupations on plantations as slaves. These slaves might gain insight and information from things they observe or comments they overhear, intelligence helpful to the Union military.

For example, the orders from the Jacksonville command to Lt. Col. Wilcoxson in St. Augustine to seize a large shipment of cotton from Braddock's Farm were likely inspired by intelligence obtained from scouts.

---

2. There were some White scouts, typically Confederate Army deserters. Raising a "poor man's army" as the South did, many conscripts realized that they were fighting for wealthy plantation owners not required to serve and whose well-being was not in jeopardy. Conscripts were unable to provide for their family and often did not get paid. They also struggled with meager rations, inadequate basic supplies, and harsh discipline from officers. Desertion was a problem among both armies, but particularly acute among the Confederates and in Florida where large encampments of deserters would battle with regular army trying to capture them. Half of the Union force landing at Cedar Key in February 1865 were Confederate deserters who had enlisted in the Union army (2nd FL Cavalry); the other half comprised freed or runaway slaves (2nd FL Infantry). See "Blockaders, Refugees, & Contrabands: Civil War on Florida's Gulf Coast, 1861-1865" by George E. Buker (1993: University of Alabama Press) for more.

As Capt. Dickison had his informants throughout the White population, the Union used its scouts to gain intelligence from informants in the Black enslaved population.

We should also remember that this is 1865 and the interior of Florida is sparsely settled – still largely untrammeled wilderness – and its terrain was frequently mapped unreliably or unmapped altogether. Scouts who are familiar with the geography, or who can gain familiarity with the geography through their secretive movements, would have been very helpful in Union operations.

Such scouts may have been operating in a manner akin to Harriet Tubman who was also referred to as a "scout." They may have been aiding escaping slaves to get to safety either overland to the Union garrison at St. Augustine or north on the St. Johns River to Union-occupied Jacksonville. In either case, the scouts on the raider team could have been quite familiar with navigating the St. Johns River in the dark, remaining undetected, and experienced with the preferred routes to go on foot to St. Augustine, both key aspects of the raider's movements for their mission.

Yet having seven scouts seems like a large number to act merely in a guidance role through the Florida scrub. Could these men be more than "scouts"?

There were reports of small units of self-organized former slaves who conducted guerilla warfare outside of Union military operations in Florida and elsewhere. It seems likely that, given the number involved, these seven "scouts" were in fact a guerilla group unto themselves who were included in the raid team. In the Appendix, there are references to Black scouts acting in guerilla operations in Florida.

Sergeant Harmon's letter will identify some "scouts" by name: Israel Hall as "chief scout" and Henry Brown as "scout" as well as naming "Ben. Gant" who is oddly referred to as a "citizen," but is likely also a "scout." More on this later.

With 16 from the 3rd USCT and 6 from the 34th USCT plus 7 "scouts" or guerillas – 29 men – we have one more raider team member to identify. This one is perplexing.

According to Harmon's letter, this individual comes from the 107th Ohio Volunteer Infantry (OVI). This unit was called 'the German regiment.' The 107th OVI was a White unit, making this raider a White soldier. Clearly, he is not an officer, just a soldier.

Research indicates that the 107th OVI left Jacksonville and was sent to Hilton Head Island, South Carolina for the fight for Charleston in December 1864 and was a regiment occupying Charleston in March 1865. It was unlikely that anyone from the 107th OVI would have been available in Jacksonville in March for the raid.

However, the 75th OVI remained in Jacksonville. Maybe the White soldier from the 75th OVI was German-speaking or had a German surname, fostering the belief that the White soldier was from 'the German regiment,' the 107th OVI.

Why does this lone White soldier become a member of the raider team? We don't know and he is never mentioned again in Sgt. Harmon's letter. That leaves us to speculate.

Given the risky nature of the mission and its timing as the war is about to end, the soldiers agreeing to serve on this mission had their reasons and their purposes. Adding a White soldier to an otherwise all-Black group must have had a defined purpose.

The 75th OVI (and 107th OVI) had plenty of experience in operations together with Black soldiers in Florida, so working with Black soldiers would have been nothing new. Still, this addition to the raider team is curious.

It suggests that this odd White soldier had some specialized function that was necessary for the success of the operation. Considering what the raid involved, there was nothing specialized about paddling a pontoon boat, burning a set of buildings and/or a bridge, raiding a plantation, freeing slaves, or being experienced in combat. The other raiders had had some prior experience in such operational activities.

The only specialized task that sticks out is the operation of a flatboat, particularly when loaded with potentially difficult cargo like wagons, horses, and mules. If there is trouble in managing the flatboat, the operation could be hamstrung. The raid leaders likely knew that this flatboat operation was not something to take for granted. Also, perhaps their superior officers insisted on an experienced flatboat pilot to approve the mission.

Let us imagine how this may have evolved.

Seeking an experienced soldier with solid experience managing and piloting a flatboat on a river carrying wagons, mules, and horses, the raiders may have first sought out a Black soldier with the requisite experience among the hundreds, likely over a thousand Black soldiers based in Jacksonville. Not only would this Black soldier need to be an experienced flatboat pilot, but this soldier also had to be willing to join a crazy risky mission deep into enemy territory with the war nearly over. It seems they didn't get anyone both experienced and willing from among the Black soldiers.

The raiders ought not to leave this key piece of the mission to chance; this role of flatboat pilot must be filled by someone qualified. They turn to the White soldiers or perhaps are referred to possible leads by commanding officers.

We should note that Major Frederick Bardwell of the 3rd USCT was a math professor at Antioch College in Ohio prior to the war. His first enlistment was as a Private in the 2nd Ohio, and then was later commissioned a Second Lieutenant in the 10th Light Artillery [3]. Bardwell may have had former students or army friends in the Ohio regiments stationed in Jacksonville, and he may have been the connection to the White Ohioan on the raider team

Going to a White soldier adds another qualification besides flatboat experience and willingness to join the raider team. The White soldier will need to be able to take orders from a Black commander.

---

3. From an undated biographical article: "Frederick William Bardwell - Major of the 3rd U. S. Colored Regiment - The Case of the Gallant Professor" at the University of Kansas website under Student Affairs-KU Memorial Union - https://union.ku.edu/case-gallant-professor

This goes to a key reason Black soldiers were not commissioned officers commanding operations; White soldiers could end up refusing an order from a Black officer in a difficult situation.

At this point in the evolution of the war, Black soldiers had been fighting alongside White soldiers for nearly two years, including the 75th OVI and 107th OVI. While there had been mutual respect earned, extending that respect to the point of White soldiers obeying leadership from Black soldiers was not anything that had been practiced or experienced by White soldiers.

Perhaps the raiders were fortunate to find a White soldier motivated by the mission and ready to join the team, having no problems with taking orders from a Black superior non-com officer as operational mission commander. On the other hand, perhaps they found someone who was well-qualified, but who needed some incentive, like a monetary or another enticement, to take on this role. We cannot know.

In any case, they added a White soldier to the raider team, presumably to fulfill a key role like piloting the flatboat.

If the White soldier was the flatboat operator and chosen specifically to fill that role, it may explain why there is no mention of either the White soldier or the flatboat in Sgt. Harmon's description of the action. Since Sgt. Harmon's letter was for a Black audience and was descriptive of the achievements of the Black raiders, his omission of any further reference to the White soldier or his possible role with the flatboat, and even the operation of a flatboat during that segment of the mission would be understandable.

The contributions of the White soldier were not at all what the letter and its narrative were about. It was about the achievement of a Black-planned operation and a Black-led mission of Black soldiers and that was letter writer Sgt. Harmon's primary focus.

The raider team is assembled. They have the command's approval. Let the mission begin.

Chapter Nine

The First Leg of the Journey

March 7-8

Sergeant Harmon describes the start of the journey:

> *After waiting some time for darkness to throw her pall over the scene, the commander gave the order to push off. The party then moved up the St. John's River, in pontoon boats to Orange Mills, where he [the commander] landed with ten men and skirmished the country to a point near Pilatkia, where the boats met them ....*

Sergeant Harmon does not break things down day-by-day but notes that they started from Jacksonville when darkness fell on March 7. We know that the raid occurred on the evening of March 10 and that they arrived in St. Augustine sometime on March 12. Therefore, our breakdown of dates and times is theoretical.

Also, non-Floridians should note that the St. Johns River is one of the largest rivers flowing from south to north (see the Nile River in Egypt for a comparable example). When moving "up" the St. Johns River, one is heading south, and of course when moving "down" the river, one is heading north.

The raiders are traveling in three pontoon boats which each hold 10-12 individuals typically.

"The commander" is Sergeant-Major James who has the boats land just north of Palatka (Harmon spells it "Pilatkia" as it is on Union maps) at Orange Mills.

Civil War-era pontoon boat

Did they travel 50 miles south by pontoon boat in one evening? Sergeant Harmon does not provide a time frame for their movements, but it is certainly possible.

Disembarking at Orange Mills north of Palatka, Sergeant-Major James takes ten men to "skirmish" the area around Palatka, presumably from the relative safety of the eastern side of the river, the western side being the location where Dickison and his men could be located.

That seems to be the intent of this maneuver, to scout for activity on the opposite shore that might indicate the presence of Dickison's forces. The raiders are relying on Dickison being at his camp in the vicinity of Waldo. If Dickison's men are present in Palatka, that would put a severe crimp in the mission plans.

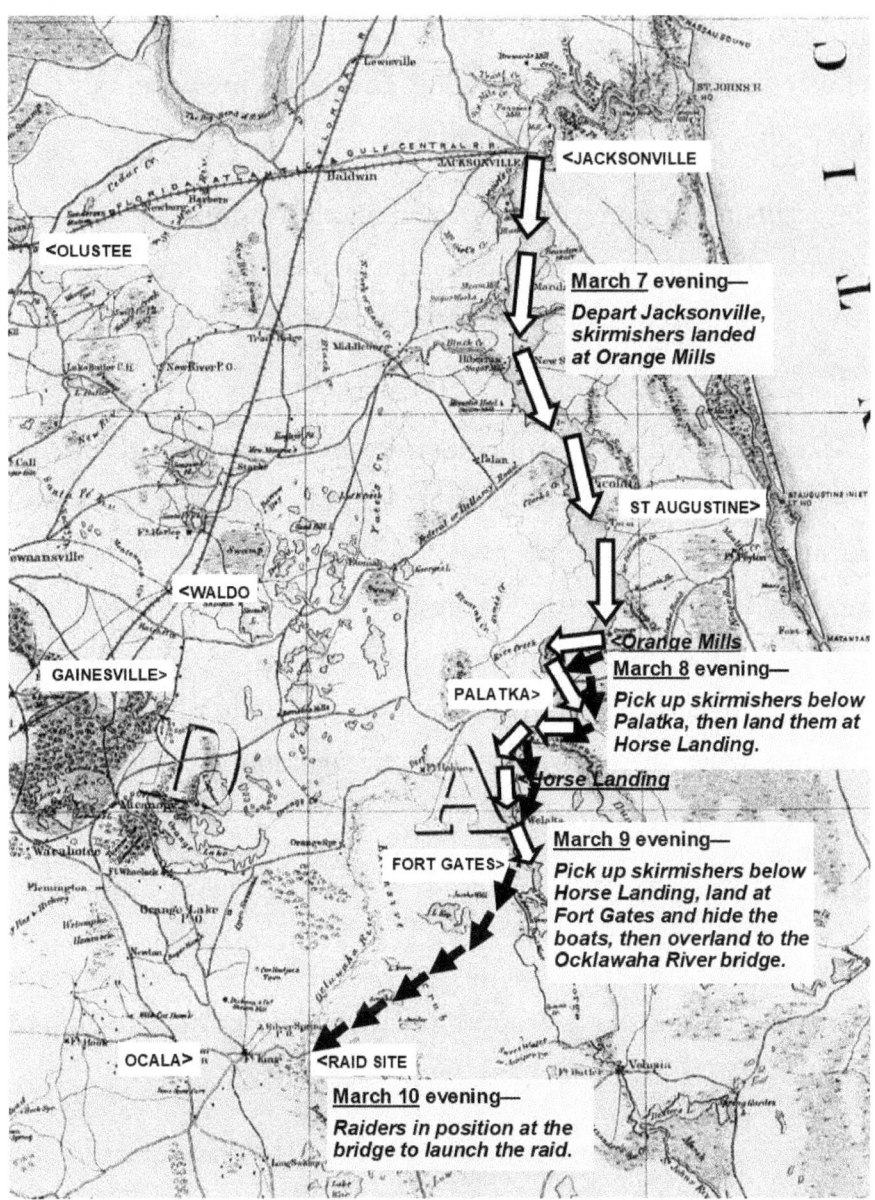

Sergeant Harmon uses the word "skirmish" rather differently from our common use. Reading "skirmish," we expect there is a small encounter with a body of opposing combatants. Akin to the vernacular of the day, Harmon uses the word "skirmish" in the sense of seeking out a body of opposing combatants rather than encountering them necessarily. We would find "reconnaissance" to be more apt since clearly the raiders were looking for Dickison's presence, but they were not seeking any direct encounter.

Sergeant-Major James and ten men would reconnoiter opposite Palatka, watching for movement and activity in daytime hours after finding spots to rest and to observe. A telescope or spyglass would have great value, both now and later.

Dickison and others describe the town as deserted due to the frequent Union occupations. If James and his skirmishers see movements that suggest trouble, they would return to where the boats were hidden, and where the remainder of the raider team were resting nearby during the day, and then figure out what to do. If there was nothing amiss in Palatka, James's team would simply wait for darkness when the boats would resume their journey south on the river and pick them up.

March 8-9

The evening of March 8 has the boats arriving to pick up the "skirmishers." Then they are dropped off again a short distance south of Palatka near Horse Landing to skirmish that area.

You will recall that Horse Landing (mislabeled by Harmon as "Horse Shoe Landing" even though it is an accurate description of the shape of the river at the location) is where the *USS Columbine* met its fate amid a Dickison ambush. The sunken, charred hulk of the *Columbine* remains at Horse Landing.

James and his skirmishers are going to watch this location, too, as the only other likely location where Dickison's forces might have a presence around the landing area.

March 9-10

Around 9 pm, James's team is picked up and the three pontoons travel another short hop to Fort Gates where the boats are easily hidden away in the dense swamp nearby.

Postcard: wild banks of the St. Johns River

We can imagine that the scout(s) who knows where the flatboat is hidden has checked on its location to ensure that it remains where it is expected to be. The scout(s) as well as the White soldier who may be the flatboat pilot may have stayed behind to guard it.

The raiders proceeded southwest through the scrub toward the Ocklawaha River and the bridge near the Marshall Plantation. They likely spent the day resting in hiding nearby since the distance between Fort Gates and the river bridge is about 30 miles, enough for one day.

Chapter Ten

The Raid

March 10 evening

Given the timeline of activity on the night when they raided the Marshall Plantation, the raiders likely started a few hours after nightfall, perhaps 7 or 8pm.

At the bridge

Chronologically, Sgt. Harmon's letter omits a detail of the raiders' arrival at the Ocklawaha River bridge, noting it later in his narrative.

In that later text, it is noted that three individuals, presumably armed, are at the bridge when the raiders arrive to cross it. The raiders would not have wanted a confrontation that would provide any kind of alert to the area. With these armed men visible on the bridge, likely toward the middle of a narrow, single-wagon width, 100-foot span, there were few options. Harmon relates: "… having to charge the bridge in going to Marshall's and killing three rebels had only stirred them up…."

1928 Sharpes Ferry Bridge across the Ocklawaha River, replaced in 2012

What was this body of armed locals doing at the bridge? (If they weren't armed, there would have been no need for charging and shooting them.)

They are named "rebels," a term that designates any who are set to oppose the presence and work of the Union raiders, not necessarily soldiers or militia.

By the way, the use of the Confederate gray uniforms was mostly present among the armies of Virginia and Tennessee. Confederate soldiers in other areas, Florida in particular, wore everyday clothes, making soldiers quite indistinguishable from non-military folks bearing arms. Confederate officers were most likely to wear the gray uniform.

Harmon's designation of them as "rebels" is common in Union accounts of armed resistance in Florida (and elsewhere) and indicates that these armed men on the bridge were no different in appearance from other Confederate soldiers dressed in everyday clothing that had been encountered elsewhere.

Given their location at the bridge, on a moonlit evening, carrying rifles, it seems they were at that place for hunting, watching for deer or other game coming to the riverbanks. Using the view of the river afforded by the bridge makes perfect sense for hunters.

The men on the bridge are likely looking out on the river at its banks, not at the end of the bridge when the raiders approach.

Harmon says the raiders had "to charge" the bridge. Upon mutual discovery, both sides likely acted quickly against the other. By the time the surprised group of locals turned to present their rifles to combat the identifiable Union soldiers, the raiders were ready with shouldered rifles and opened fire. The narrowness of the bridge allowed the soldiers to concentrate their fire lethally. All three of the locals were killed. None seems to have escaped.

Raiding the Marshall Plantation

The raiders proceed to the Marshall Plantation which had not apparently been alerted or at least alarmed by the gunfire at the bridge, assuming that the noise could be heard at a distance. Indeed, if the men on the bridge were hunters from the Marshall Plantation, the sounds of gunfire might have been expected and dismissed as hunters' shots. No one would have expected Union raiders in southeast Marion County.

We have no idea how many defenders were at the plantation to provide any resistance to the raiders. Sgt. Harmon does not mention any resistance.

Here is Sgt. Harmon's account:

> *Here the expedition captured some 25 horses and mules, burnt a sugar mill, with 85 barrels of sugar, about 300 barrels of syrup, a whiskey distillery, with a large amount of whiskey and rice, and started on their return, bringing along 95 colored persons, men, women and children, re-crossed the Oclawaha River, burning the bridge.*

Photo of a sugar plantation operation

Again, we can hardly imagine that Sgt. Harmon or any of the raiders stood around counting all the barrels of goods discovered at the plantation, and he allows for an estimation of the number of horses and mules captured. Yet even Dickison's account admits that there were "200 hogsheads of sugar" destroyed. (A hogshead was twice the size of the standard barrel, likely 55-62 gallons.)

Regardless of any estimates from either side, the Marshall Plantation had plenty of finished products and resources apparently ready for shipment, and the raiders made sure to destroy all of it. Given the number of stores on hand, plus the additional slave labor, this may have been a time of year of high production.

A curious matter is the number of slaves found on the plantation. The 1860 Slave Census for Marion County shows two entries:

MARSHAL, J. Foster, 46 slaves

MARSHALL, J. F., 41 slaves

The Marshalls of South Carolina had another income plantation in northern Marion County at Wetumpka which produced cotton. Presumably, the Slave Census listing reflects two different plantations listed for the same owner, Col. Jehu Foster Marshall. What is striking is the difference in numbers between the accounting on the census compared to Sgt. Harmon's claim that 95 slaves were taken from the sugar plantation, a number closely mirrored by the Union report claiming the raiders took 91 slaves from the Marshall Plantation.

Sugar cane grinding in 1928

Both cotton and sugar plantations required intense, time-sensitive labor and had roughly matching seasons with harvests in the Fall and planting in the Spring. While cotton needed to be plucked soon after the boll opened, since a mere rainstorm could ruin the boll's contents, sugar cane also needed to be processed quickly to prevent the loss of its sugary content.

For slaves from the cotton plantation to be brought to the sugar plantation seems logical, taking for granted the honesty of the reporting on the Slave Census. Perhaps the further refinement of the Fall's sugar cane harvest, taking syrup to sugar and distilling whiskey, has brought about this increase in slave laborers at the southeast Marion sugar plantation. Planting preparations may also have been a reason for the large influx.

This notion of engaging in further refinement of the sugar cane product may explain why hundreds of hogsheads of syrup are still present when the raiders arrive since throughout the Confederacy there was a desperate need for any and all resources by 1865.

With the Slave Census registering numbers from 1860, a great deal of turmoil had occurred in the state in the intervening five years to the time of the raid in March 1865.

Thousands of Florida slaves had escaped or been liberated to Union army sanctuaries, depleting the slave labor force anywhere near Union forces whose presence acted like a magnet for those seeking freedom. Even slaves in Marion County sought passage to Palatka during several Union occupations, or simply navigated the St. Johns River north to Jacksonville, or went overland through the wild scrub to the sanctuary of the Union garrison in St. Augustine. One report claimed several Marion County slaves seeking to escape by boat on the St. Johns River were lost and presumed dead when their capsized boat was recovered.

Further, with Union disruptions in the taking and abandoning of Jacksonville on three occasions prior to the fourth and last occupation in February 1864, many slave-holding Whites in northeast Florida fled further and further into the interior, abandoning their farms but bringing their valuable "property" in slaves. Selling, leasing, or renting these slaves to ongoing plantation operations, particularly to meet peak seasonal demands, provided the owner with some funds to survive.

These scenarios make it easy to see how the overseers at Marshall's plantations may have brought in more slave labor by March 1865 than what was listed on the 1860 Slave Census, likely as a supplemental slave labor force during a particular season in one of the operations.

The raiders are hustling 95 slaves, "men, women, and children," out of the plantation along with several prisoners from the White plantation staff together with wagons, horses, and mules. This now large body of people, over 120 total, plus animals and wagons heads east for about a mile to the Ocklawaha River bridge. Once crossed, the bridge is also set afire.

With a complement of soldiers from the 34th USCT (former 2nd SCVI) and their training in "jayhawker" tactics, the fires set at the plantation and the bridge are likely their handiwork. The plantation fires were so thoroughly devastating that no agricultural operation would ever again occur there.

There are four prisoners mentioned at the end of Harmon's action narrative but nowhere within Harmon's account is it disclosed when these prisoners were taken. The best guess is that they were all captured at the Marshall Plantation.

We should also reckon with the number of plantation staff discernible since there were over 90 slaves engaged in work on the plantation. Having taken four prisoners, we would imagine that more than four staff would be needed to provide oversight for so many slave laborers. We can add the three men shot at the bridge as possibly being Marshall Plantation staff. That would bring the count to seven. Perhaps several may have escaped into the scrub along with one who was able to flee on horseback to alert the Ocala authorities. In any case, it seems unlikely that the four prisoners account for all of the plantation's White overseers and staff.

The Alarm Raised

Word of the raiders and their raid gets back to Ocala and the Home Guard. We don't know how it happened, but it did. This may account for another White overseer of the large slave population at Marshalls.

One local source, *Ocali Country: Kingdom of the Sun*, states that a "servant" from the plantation rode away to alert the Home Guard.[1]

Another local source, David Cook's *Historic Ocala*, more explicitly claims it was a "slave," but that may have been an interpolation of *Ocali Country's* use of the word "servant" which is often used euphemistically to refer to a slave (or "property.")

Another more raid-contemporary source, but second- or third-hand, claims a "girl" rode to Ocala. This report comes from a missionary serving in St. Augustine, Mrs. Greely, who had learned about the raid from her sources. She states the following in her report to Brother Whipple in a letter dated March 18 (see Appendix for the full text):

They would have brought more people and more booty had they not been betrayed by a girl on a plantation where they had killed the

---

1. *Ocali Country* then asserts that the Home Guard intercepted the raiders at the plantation, a physical impossibility. While the volume is an historical source, it merits questioning since it offers numerous errors about the raid.

Overseer, & burned the sugar mills with a quantity of sugar syrup & whiskey and the body of the Overseer in the sugar house.

Continuing with Mrs. Greely's note, there is no mention of killing anyone at the Marshall Plantation by Sgt. Harmon or other sources. Would he have omitted this detail? That seems unlikely, but the gruesome description by Mrs. Greely may have been disturbing for Harmon's readers. Such conduct would not be incompatible with the "jayhawker" style that Col. Montgomery had used when commanding the 34th USCT (2nd SCVI). However, it also seems out of place for the raiders to kill the overseer unless there was some personal matter to settle. This recalls the previous discussion about the six men from the 34th USCT possibly coming from Marion County and even the Marshall Plantation. No other account among the sources mentions the killing of an overseer.

Overall, it seems this account in Mrs. Greely's letter is simply another elaboration of the story, one likely without merit since it is joined to other unlikely or inaccurate details in Mrs. Greely's letter.

Neither Sgt. Harmon nor Capt. Dickison makes any mention of how word got to Ocala.

It seems more likely that a White member of the plantation's oversight crew played the messenger-on-horseback role, but admittedly it could have been a slave, even a "servant girl."

The Confederacy apologist's agenda of portraying plantation life as a pleasant experience for Black slaves who served their masters willingly and gratefully could be at work in a claim like this, that a slave was alarmed by the destructive, invasive acts of the raiders, was opposed to Union military efforts, and was supportive of the Confederacy of the master. A slave behaving in this manner seems unlikely but is not unheard of.

Holly Plantation ambush

After crossing the Ocklawaha River and setting the bridge on fire, Sergeant Harmon's letter tells how a detachment of soldiers under Sergeant Joel Benn of Company B, 3rd USCT was sent to the nearby Holly Plantation (which Harmon misspells as "Hawley").

The Holly Plantation had 27 slaves according to the 1860 slave census, suggesting that it was not a very big target.

This action seems unlikely for a planned strike. The raiders already had more than they could handle with over 90 liberated slaves and a handful of prisoners plus wagons, horses, and mules. They didn't need any more to manage and time was of the essence. Why this seemingly off-script excursion? Or was it part of the original plan?

Sergeant Harmon presents the Holly Plantation probe in a matter-of-fact manner. In his letter, he cites the Marshall Plantation as "one of the objects of the expedition." Does this mean that the plan was to raid Marshall's, burn the bridge, and then raid the Holly Plantation? Here is the text from Sgt. Harmon's letter detailing the raiders' movement up to the beginning of the Marshall raid:

> ... *proceeded with the whole force across the country to the Oclawaha River, to what is known as Marshall's plantation. Here was one of the objects of the expedition ....*

The reader cannot tell if, among the "objects" of the expedition, there was more than the Marshall Plantation to be raided, like the Holly Plantation, or if Sgt. Harmon was referring to other 'objectives,' like the objective of moving undetected from Jacksonville to the point in the letter's narrative at the Marshall Plantation as well as the objective of completing the journey to St. Augustine's safety after the raid.

Sgt. Harmon's account does not indicate why the Holly Plantation was a location to be engaged, suggesting that even Harmon's first-hand source may not know the reason for the Holly Plantation mission.

Frankly, the mission is inexplicable. The modest Holly Plantation seems a highly unnecessary diversion. A decent explanation for the foray begs to be offered.

One method of getting some insight into this operation is to consider who *was* present and who *was not* present in the 9-man raider contingent sent to the Holly Plantation.

We are eventually told that six soldiers from the 3rd USCT led by Sgt. Joel Benn accompanied at least three scouts which we learn later included the "Chief Scout" Israel Hall.

We know that the six soldiers from the 34th USCT were experienced "jayhawker"-style combatants. If the incursion intended to destroy the plantation and liberate the slaves - a repeat of the Marshall raid - one would have expected the six men of the 34th USCT to be assigned to the contingent rather than six men of the 3rd USCT. This absence of the 34th USCT suggests that a repetition of the

Marshall raid with the torching of the whole Holly operation was not planned.

Then what was the plan?

Interestingly, the Chief Scout and two other scouts are part of the Holly contingent. The presence of the Chief Scout would indicate that this was part of the scouts' agenda and that the 3rd USCT unit was accompanying the scouts for "muscle" if needed.

What possible reason could the scouts have for generating this side mission to the Holly Plantation?

Remember that the scouts likely received critical information making the raid possible - the flatboat location plus the large influx of slaves at the Marshall Plantation. The scouts' likely source was someone on a nearby plantation, either Marshall's or another, like the nearby Holly Plantation.

If an informant slave had lined up information for the scouts that pointed to a target at the Marshall Plantation but was enslaved at another plantation, it would be reasonable to imagine that the informant slave would strike a bargain for providing the information. Knowing that a slave-liberating raid would be nearby, the provision of the intelligence by the informant slave may have been predicated on their extraction from the Holly Plantation during the raid operation.

The best guess for the Holly Plantation foray is that it was an *extraction operation* led by the scouts and supported by Sgt. Benn and his men from the 3rd USCT. This was likely in fulfillment of a promise made by the scouts to an informant slave on the Holly Plantation that this person (or persons) would be extracted and freed when the Marshall raid occurred.

In any case, here is Sergeant Harmon's description of how things unfolded:

> *Six men then were detached from the command and sent under charge of Sergeant Joel Benn, of Co. B, 3d U.S.C.T., with Israel Hall, scout, to Hawley* [sic] *plantation ....*

Parsing Harmon's words, the six men "detached from the command" would likely indicate soldiers within Sgt. Benn's unit. There was also a scout, who it seems had brought along two other scouts, making nine altogether, nearly one-third of the raider team.

Likely alerted by either the gunfire at the bridge or by the fires set on the Marshall Plantation and/or the bridge, those responsible for the Holly Plantation seem to have been ready for any trouble that might come their way. The plantation's defenders likely positioned themselves in the dark shadows out of the moonlight, watching and waiting to see if any raiders would venture their way.

As the 9-man unit advanced, they were surely unaware of the danger that awaited. Suddenly, armed defenders of the plantation emerge from the shadows to capture the two scouts and then start shooting at the soldier detachment watching nearby. As gunfire erupts around them, the toll is high as Sgt. Harmon continues his account:

> *... and they were attacked by a small body of rebels, and Sergt. Benn was killed, shot through the heart, Henry Brown, scout, wounded, and Israel Hall, chief scout, captured, as was another citizen named Ben. Gant, the others being compelled to return to the main body.*

Two of the raider team's leaders are casualties– Sergeant Joel Benn is killed and the chief scout is captured along with another scout while another scout is wounded.² This is quite a blow to the small raider unit.

---

2. There is a second-hand (or more) source in a letter from the Clerk of Company A of the 3rd USCT – the raiders were from Company B – published in The Christian Recorder on April 29, 1865, one week after Sgt. Harmon's letter is published, discussing the Marshall Plantation raid. The letter is of poor quality as a source, presenting numerous inaccuracies. Regarding Sgt. Benn: "when he received his death he was endeavoring to shield a helpless woman and her child from the hands of those God-forsaken traitors." This may have been an attempt at hallowing the loss of Sgt. Benn with a tale of gallantry. It seems advisable to continue the belief that the detachment to the Holly Plantation was ambushed and likely was unable to accomplish anything besides having the surviving members of the raider detachment flee to safety with the main body. Had there been any smidgeon of success (or glory) derived from this foray, Sgt. Harmon would have noted it, but there is nothing.

Sergeant Major James must have authorized this side venture to occur, dispatching Sgt. Benn's unit to provide support.

In the Holly Plantation account, we find Harmon mixing his terminology, referring to Ben Gant as "another citizen" when he had already referred to "Henry Brown, scout, wounded, and Israel Hall, chief scout, captured." Why didn't he refer to Ben Gant as a scout? Given his use of the term "citizen" to label the seven scouts on the raider team at the beginning of his letter, we would do best to assume that Gant, too, was a scout, captured along with Israel Hall, the chief scout and presumably the wounded Henry Brown as well. (Later, describing the fight in the scrub, Harmon will refer to 'the sharp crack of the soldiers' rifle and the louder roar of the citizens' fowling-piece,' again using "citizen" to describe one scout's fowler, a hunting rifle distinctly louder than the sharp crack of standard issue Union muskets.)

Local lore has provided divergent accounts of the Holly Plantation encounter which include one Holly son being captured and later escaping[3], and another[4] claiming the home was partly set ablaze. It seems that the real story was lost to the Holly family which has instead repeated unfounded claims. It is too bad that the family has been unaware that their ancestors handily dispatched a team of Union raiders, killing a Black Union sergeant and capturing two or three Black scouts in repelling the intruders and defending the homestead. The real history is superior to the lore!

Again, one would expect that Harmon would mention any benefit of this unfortunate excursion like the taking of a captive or the freeing of slaves (or a valorous act by Sgt. Benn), but nothing like that is present in his dismal account of that operation.

Further, the description reflects an ambush – "they were attacked by a small body of rebels," states Harmon's letter, not that the detachment did any attacking, destroyed anything, took anyone prisoner,

---

3. Local lore provided in *Ocali Country: Kingdom of the Sun* (an account replicated in Cook's *Historic Ocala*) states that these Holly Plantation raiders took one of the Holly sons, Franklin, as a prisoner and that he later escaped into the scrub. It is irrelevant in any case, but Harmon mentions no captives being taken.

4. Some Holly family lore was preserved in the interview notes of local historian Scott Mitchell, and in the pamphlet named *Marion County Remembered*, specifically "Salty Crackers" Number Four by Sybil Browne Bray, Fall, 1986. In those accounts, the main home of the Holly family was set ablaze which again seems unlikely.

or freed any slaves. An ambush presents an unlikely scenario for capturing anyone and bringing them back to the raider caravan amid a humbling retreat.

The misfortune of the Holly Plantation ambush meant Sergeant-Major James would focus the plan going forward on their movement toward St. Augustine, now moving with haste, having had one of the team's leaders killed and another leader captured, plus another "scout" captured and yet another wounded and likely captured. With these casualties on his mind, Sgt-Major James pushes the caravan of people and wagons forward, commencing its journey through the scrub toward the river crossing at Fort Gates nearly 30 miles away.

Chapter Eleven

The Chase Begins

Dickison learns about the raid

Confederate cavalry Captain Dickison's report made a note of his first communication regarding the raid (it was recorded in his report on March 15), but we should question this recollection:

> *On the evening of the 10th inst., I received information from Marion County, through Col. Samuel Owens* [commander of the Ocala Home Guard], *that the enemy was advancing by way of Marshall's bridge and had advanced 12 miles into the interior, burning the bridge.*

[Note: "inst." is an abbreviation for *instante mense*, meaning a date of the current month, such as "the 5th inst." The abbreviation was commonly used at that time.]

A few things seem odd about this recollection of what is supposedly the first communication.

The message is received "On the evening of the 10th," a point in time when scant information about the raid could have been available for communication from Ocala.

Dickison's account states that he was informed about the burning of the bridge and a 12-mile advance by the raiders into the scrub, also an indication that the raiders were most likely headed for Fort Gates to get across the St. Johns River. None of this would have been known prior to the time when Dickison would have left his camp near Waldo to head to Marion County, likely 10-11pm.

It is most likely that Dickison is referring to information provided via courier when he gets near Silver Springs around dawn on the 11th.

If he had received this information in the initial communication, he would have understood what was happening and been able to figure out the raiders' plans. He would not have ridden to Marion County but headed to Palatka instead to get across the river and intercept the raiders.

The nighttime journey that brings Dickison's cavalry near Silver Springs at dawn reveals his lack of knowledge about what is happening in Marion County. Most likely, Dickison knows only that Union raiders were attacking in Marion County, and is aware that the local militia would be unlikely to deal with them alone. Dickison makes haste for Marion County.

Dickison writes:

I immediately ordered out my command and in two hours was in rapid march in that direction.

At what time might Dickison have received this initial communication?

Somewhere in the time frame of 7pm to 8pm is when the raid may have begun. A rider escapes the plantation and travels 8-10 miles to Ocala, a half-hour or so ride from the plantation. Word of the raid would need to be delivered to the Home Guard's leader Colonel Owens (or another leader) who then gets a telegraph operator to dispatch the message.

By 9pm or so, having located someone with authority like Col. Owens, and then a telegraph operator, the report of Union raiders at the Marshall Plantation would have been wired and delivered to Dickison in Waldo. Two hours later according to Dickison, likely before midnight, the Captain and fifty of his cavalrymen are moving quickly toward Marion County.

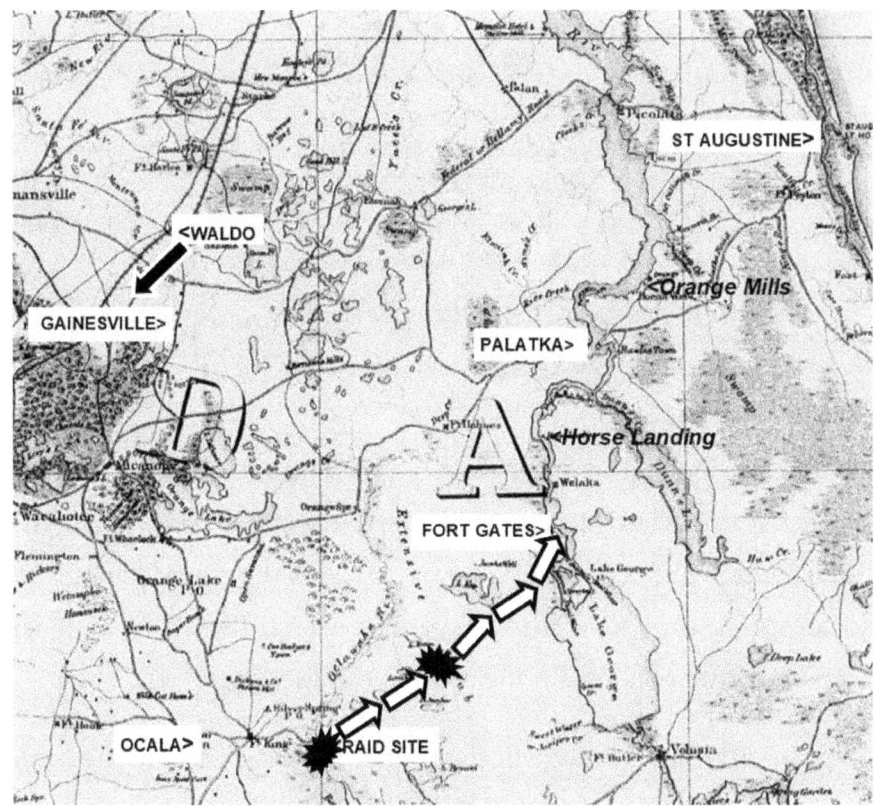
Movements to midnight March 10-11

Gunfight in the scrub

After sending the telegraph message to Dickison, Col. Owens authorizes someone like Captain Howes to quickly round up anyone available as Home Guard defenders. (Col. Owens is quite old and weak at this point, but he and his family carry considerable weight in the community. He would not have been part of the responding Home Guard.) Howes probably has reliable men he seeks out for the mission and picks up anyone else who has a rifle and a horse and will volunteer.

This is the end of the war. It has taken its toll on the White men in Ocala and Marion County. There is a report noted in *Ocali Country* (p. 84) that states:

> *By midsummer [1864] even home guard units had been withdrawn for service elsewhere. Boys under age, and older men excused by ill-health, were enlisting.*

Ocali Country later notes (p. 85) that recruits to assist Dickison's defense against the Union incursion into Alachua County in August 1864 included 15 boys under 16 years of age for his force. What is left to constitute the Ocala Home Guard in March 1865 is likely older men and boys with few having combat experience.

Sergeant Harmon's letter will claim that there were fifty in the complement of the Home Guard. This is extremely unlikely. A local source indicates that 16 men joined with Captain Howes to address the raiders' threat. This seems far more plausible. Ocala had a population of only several hundred from which a Home Guard response could be quickly formed.

Not knowing where the Union raiders were headed or what they intended to do, the hastily assembled Home Guard simply heads east toward the plantation and the bridge, starting perhaps by 9 pm.

Again, it is about a half-hour ride from Ocala to the bridge. When they arrive around 9:30 pm, the bridge may still be burning and is impassable. It is easy to figure out where the raiders were headed.

The burning of the bridge means the raiders were not going to use it again. The raiders had gone east into the scrub.

A caravan of over 120 people plus wagons and horses is not hard to track through the sandy wagon paths of the lightly traveled scrub, even at night. The Home Guard fords the Ocklawaha River and commences their pursuit.

The Home Guard comes upon the raider caravan in another half-hour or so, roughly 10 pm.

The only account of this engagement in the scrub comes from Sgt. Harmon's letter which spends considerable time and detail explaining the action:

> *But when within about twenty miles of the St. John's River, the enemy numbering about fifty men well-mounted, came down on them [the raiders], calling on them to surrender or suffer themselves to be hanged.*

[At this point, any concerns of Sergeant-Major James that this could possibly be Dickison's cavalry disappeared. Even in the dark, he could discern that these were not Confederate regulars. Besides, Dickison would have used his force more strategically.]

> *But there was another alternative which he, the enemy, did not think of, and which the Sergeant Major, who, by the*

way is not a surrendering man, resolved to take, which was to fight them awhile first.

Seeing this, the enemy prepared himself to make it warm for the little band of colored men. Breaking to the right and left under cover of a hill, they dismounted and formed their line of attack, and came over the crest of the hill, in quite an imposing array to find the little band of seventeen men, (the balance being left to guard some prisoners and the avenues of retreat,) deployed, as skirmishers to meet them, covered as much as possible by the trees. But on they came.

And every man selecting his man, when they were near enough for every man to make sure and waste no ammunition, Sergt. James gave the command to commence firing, and for awhile nothing was heard but the sharp crack of the soldiers' rifle and the louder roar of the citizens' fowling-piece, blended with the yells of their wounded and dying. The firing on the part of our men was good, as was shortly proved, for the enemy suddenly broke for their horses, when our men, leaving their cover, dashed in among them with the bayonet and clubbed guns, scattering them in every direction, leaving some 20 of their men dead and a few wounded.

> *Finding the way clear again, Sergt. James, on summing up, found the woods had afforded them such good covering that he had only two men wounded, and after taking possession of the best of their horses, (although the enemy suffered so severely, he showed himself to be no mean marksman, as numerous holes in our men's clothing amply testifies, among which, a hole through the commander's cap, caused him to withdraw his head from a dangerous position,) he again took up the line of march for the St. John's, having to abandon one wagon on the way, and soon reached the river ....*

Again, the estimation of the size of the Home Guard unit facing the raiders is surely overstated. Sixteen to seventeen Home Guardsmen seems more accurate. Of course, being on horseback and moving amid the murk of moonlit darkness in the dense scrub, they may easily have appeared to be of a greater number.

Seventeen from the raider team were detached from the caravan to confront the Home Guard, leaving a skeleton crew to guard the rear and keep the caravan moving toward Fort Gates.

Note the first mention of prisoners having been taken in the parenthetical remarks in the second paragraph of Sgt. Harmon's letter above. We are not told when or where they were taken, but as stated previously, they must have been taken prisoner from the Marshall Plantation.

The challenge issued by the Home Guard for the raiders to surrender may indicate their naivete in experiencing a battle or in encountering Black Union soldiers, perhaps believing that they had superior numbers while also being unaware of the number of combatants they were facing in the darkness of the scrub. If they had not encountered Black Union soldiers previously and based their expectations on how slaves responded when threatened, they did not understand who they were dealing with. Whatever the Home Guard's assumptions, those notions would quickly be undone.

Using whatever trees were substantial enough to provide cover, the Union soldiers and some scouts – 17 in number – took their best positions and waited with disciplined patience for their opponents to approach.

The Home Guard wisely chose to dismount their horses and charge the raiders' positions on foot.

As rifle fire erupts around the charging Home Guard and the Home Guardsmen return fire, the experience of getting shot at by trained, veteran soldiers was likely stunning for the Home Guard.

Knowing that a capable rifleman can manage to fire three rounds per minute in a manner that requires completing a number of steps to reload, doing so while under fire and with smoke clouding the area, and while casualties are taken, causing distress and generating mayhem around them, the Home Guard may have become rapidly overwhelmed by the whole chaotic experience. The rate of fire from combat-trained Union soldiers targeting them from the darkness amid the trees would be disheartening. While the Union soldiers had plenty of training and experience, for many in the Home Guard, this may have been their first taste of real combat and it was daunting.

Sergeant Harmon's account indicates that the rifle fire of the raiders quickly caused the Home Guard to break ranks and flee, some for their horses but most scrambled to disappear in the shadows of the dense scrub.

The Home Guard seems to have aimed well with its initial volleys – see the dark humor about Sgt-Major James having a Home Guard bullet pass through his cap – but likely the Home Guard could not sustain its gunfire or endure amid the intensity of the raiders' well-practiced barrage.

Sergeant Harmon describes the Union soldiers charging the Home Guard as they scattered, ensuring that they had been pushed off the field of battle and deterred from further pursuit of the raiders and their caravan.

In the haste of their escape, the Home Guard left behind a number of mounts. The abandoned horses get gathered by the raiders and added to their captured goods.

Again, the numbers cited by Sgt. Harmon ought to be taken as exaggerated. His claim of "some 20 of their men dead and a few wounded" seems highly unlikely. In the confusion of battle, counts are often woefully inaccurate. Beyond the usual chaos and smoke, add the moonlit darkness amid the thick scrub and the result is that a minor fraction of 20 seems more likely.

Ocali Country states that four of the Home Guard were killed in their confrontation with the raiders, a more accurate expectation for a nighttime gunfight in the scrub. That report may reflect the mortal nature of one or more of those wounded.

The ultimate mortality of wounds received seems to have been the case for one of the wounded raiders who later died, Pvt. Francis Fisher[1]. And possibly another raider[2] may have later died.

Other local sources including Dickison put the number at two or three of the Home Guard dead and one or two wounded.

In any case, it seems certain that there were five or fewer total casualties among the Home Guard and two wounded (one or two of whom later died) among the raiders.

Again, Mrs. Greely's letter portrays the encounter much differently and is once again of doubtful accuracy –

> *They killed the Capt. and 27 of the men, wounding eleven and capturing four whom they brought in with them, making forty three, out of seventy of the rebels, and lost*

---

1. Pvt Francis Fisher - died of "wounds received" Apr 4 1865 - is likely as one of the wounded raiders.

2. Pvt Charles Lewis: "died May 1, 1865 at Post Hospital-Jacksonville." No cause was listed. He may have finally succumbed to wounds received during the raid or subsequent infection.

of their own number only the guide who was captured. Doesn't this show Negro valor?

And they claim a little humanity, as they say they left several of the rebels so severly wounded and alone, as their companions had fled, they thought duty to go back, a few of them, and finish them. They say when the parties met they charged upon the rebels in the name of "Fort Pillow."

(It is also important to note that Capt. J. J. Dickison and his men were never accused of killing either wounded Black soldiers or captured Black soldiers. Dickison may have been a White supremacist, but he seemed well-grounded in the proper conduct of warfare.)

We can be confident that Sergeant-Major James does not want any of his team to spend much time walking around counting dead bodies. One would imagine that James sought to quickly gather the Home Guard's horses to deter their pursuit and catch up with the caravan headed for Fort Gates. There was no need to loiter and await a return of re-grouped Home Guard attackers.

Confident that they were secure for a while, raider leader James pushes everyone and everything toward Fort Gates.

It is likely past midnight by the time they have traveled the last 20 miles to the landing at Fort Gates according to Sgt. Harmon's letter. They would recover their pontoon boats from hiding and grab the flatboat from its secret location.

Their travel after crossing will likely be aided by what they have captured. Wagons and horses would provide an alternative to walking for those who are ill, injured, old, encumbered, or exhausted over the much longer haul to St. Augustine still to come. Sgt. Harmon's letter does note that one of the wagons became disabled and had to be abandoned before they got to Fort Gates.

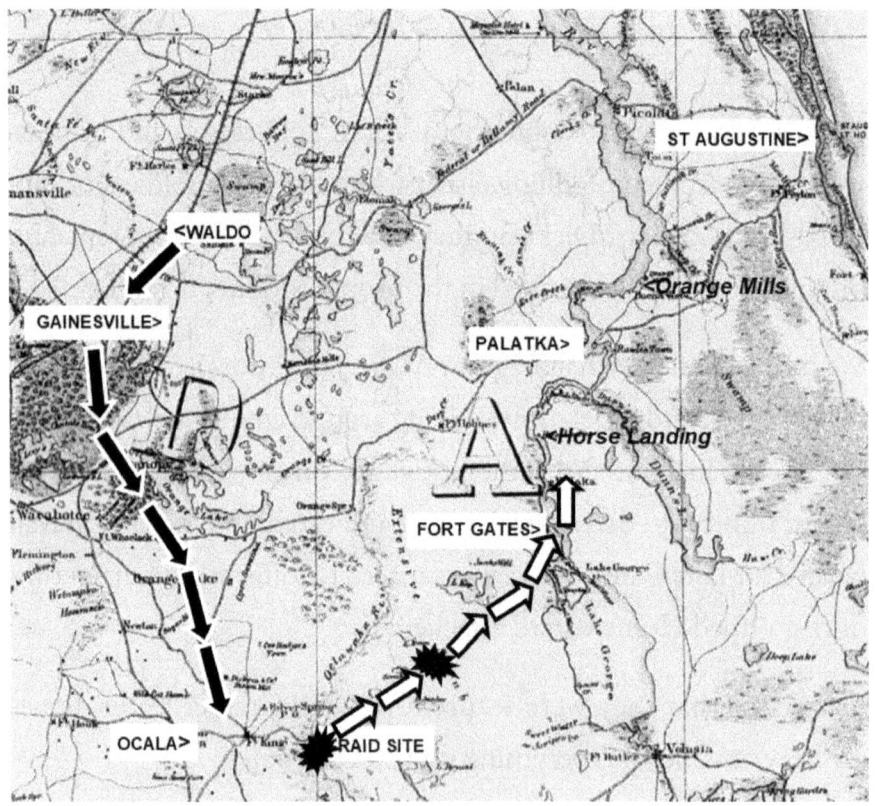

Movements to 6 am on March 11

March 11 midnight to noon

Dickison and his cavalry have a ride of about 55 miles from Waldo to Silver Springs which would take roughly 5-6 hours at a steady, reasonable pace. Dickison is upon Silver Springs as the dark of night begins lifting and a courier appears with an update, presumably from the Ocala Home Guard. From Dickison's account:

> *While near Silver Springs a courier reached me with a dispatch, stating that the enemy had burned the Ocklawaha bridge and were retreating toward the St. Johns River. I then ordered my command to march back in the direction of Palatka, and sent an advance guard to have the flatboat in readiness for us to cross the river.*

With the scant information that we believe was in the first communication at Waldo, Dickison had no idea either how large the Union raider unit was, how it had gotten so deep behind his lines, or what they planned to do. Now he is getting some clarity and the pieces are falling into place.

Also, the clever strategist Dickison is likely recognizing the strategy employed by the raiders and perceives that he has followed the raiders' preferred script quite nicely. In his mind, if not from his mouth, his preferred expletive has surely emanated. He is precisely where they want him to be and nowhere near the location where he needs to be. He is surely fit to be tied upon realizing how he took the bait. Surely furious at his geographical predicament, he finds his force now far removed from his river crossing point.

Where Dickison needs to be is Palatka and he has now been drawn 50 miles away southwest of Palatka after spending all night covering the distance. The horses are likely weary from the 55-mile journey from Waldo to Silver Springs, but he will have to work with the tired horses on which they sit and take the journey to Palatka at a slower pace. Expect the 50-mile journey to Palatka to take 5-6 hours, having him arrive there around noon.

Indeed, Dickison had not traveled far from Waldo when the raiders began ferrying people, wagons, horses, and mules across the St. Johns River. As Dickison came south to Silver Springs, the raiders' ferrying operation reached its completion, ending at roughly the same time that Dickison gets met by this messenger with an update at dawn on March 11.

The raiders have used the early morning darkness to move people, wagons, horses, and mules across the river. With three pontoon boats carrying 10-12 people each, and with over 120 people to send across the river, the pontoons will need four round trips to ferry the men on the raider team, the freed slaves, the wounded, and the captives. The same number of trips using the flatboat will bring what seems to be two wagons, plus horses and mules taken from the Marshall Plantation, plus horses left behind by the Home Guard.

Sergeant Harmon makes no mention of any issues until daybreak:

> *... commenced crossing [the river] at 12 o'clock on the night of the 10th, and at daylight on the 11th had all across except 9 horses, when the enemy coming up made [it] impossible to recross, consequently had to leave them.*

With the first suggestions of daylight, the Home Guard has regrouped and returned to follow the trail through the scrub to seek out the raiders. The caravan left an easy and quite predictable trail to follow, leading the Home Guardsmen to the river landing at Fort Gates. There they would find nine horses tied up awaiting transport by the raiders. They may also see the flatboat bringing its load across the river. The Home Guard had no way to cross the river if they wanted to, and their pursuit had to end at the riverbank. The militia would retake the nine horses and send a messenger to Ocala with another update which would take several hours to deliver, only getting to Dickison after he arrived in Palatka.

Now in relative safety on the east side of the river, it is an easy decision for Sergeant-Major James to leave behind the remaining horses still tied up on the west side. The primary task now was to move with haste toward St. Augustine, not to engage in a fight with the Home Guard over horses.

The raiders disable their pontoon boats and the flatboat (although, again, the flatboat is never mentioned by Sgt. Harmon) to ensure they aren't used again anytime soon.

While their rear flank is covered for the time being, they would have to watch ahead, not yet knowing how well Dickison had taken the bait. For all the raiders knew, Dickison's men could be lying in ambush for them anywhere ahead on the caravan's route.

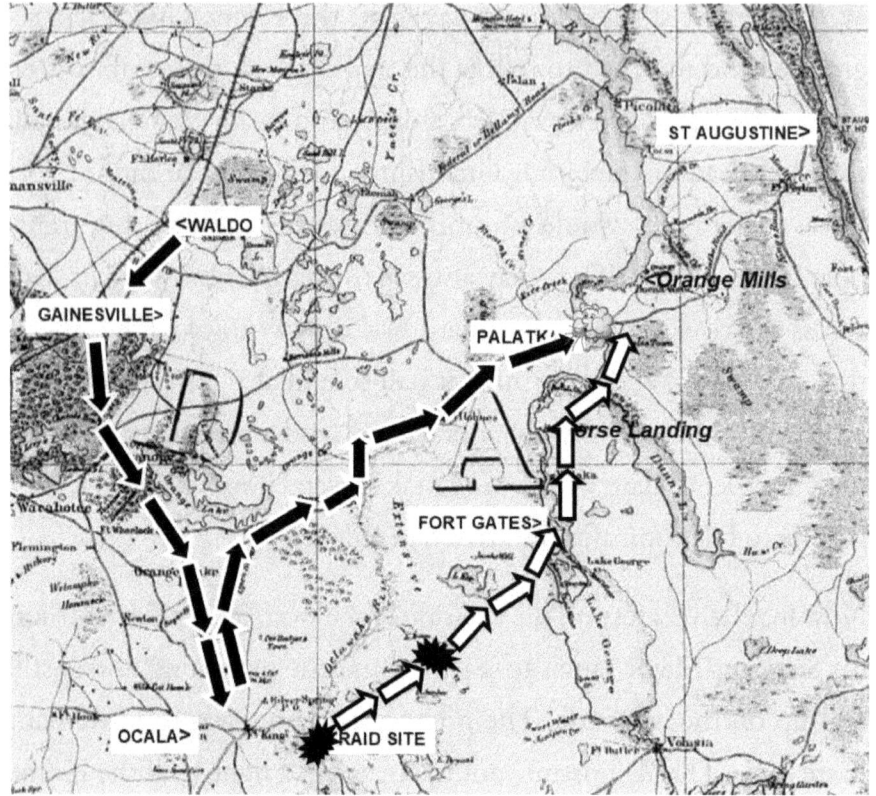

Movements to 12 noon on March 11

Moving at roughly 3 miles per hour, the caravan would have spent the morning walking nearly 20 miles north from the river landing. They would be nearing a point almost opposite Palatka on the eastern side of the river around noon, at about the same time that Dickison arrives in Palatka on the western side of the river.

March 11 afternoon to March 12

At noon on March 11, we can estimate that there is only about a mile or so separating Dickison from the Union raiders. Being on opposite sides of the river by Palatka, that mile involves the breadth of the St. Johns. Dickison must cross it to engage in his pursuit.

However, Dickison does not find favorable conditions when he arrives in Palatka:

> *On arriving at the river the wind blew very strong, which delayed our crossing about ten hours. After much difficulty, hard labor and great peril, we succeeded in crossing 50 of my command, leaving the remainder with one piece of artillery to guard and picket other points on the river. Hearing, on my arrival at Palatka, that the enemy had gone up the river in barges....*

The St. Johns River is described as "lazy," flowing at just 0.3 mph according to *Wikipedia*. However, sudden, strong windy days in March are no surprise. The mile-wide St. Johns had been made so choppy by the wind that it was impossible to safely operate a flatboat. The river can produce white caps in such conditions. It may not have been safe to cross even in a small boat if Dickison had chosen to leave the horses behind and simply ferry his men. Akin to the biblical Exodus, a wall of water prevents the enslaver/pursuer from reaching the freed slaves.

Dickison had no choice but to wait out the windstorm for some hours and then cautiously work through the turbulence in crossing the river as the wind abated.

In Palatka, his most recent update from Ocala is not at all helpful, that "the enemy had gone up the river in barges." Dickison may have scratched his head at this message, but he knew by now that his adversary was not stupid. The raiders would not be so foolish as to try to escape down the river ("up the river" would have had them headed into Lake George – that couldn't be right) past Palatka, past him; they would be sitting ducks on the river.

It was possible that a gunship was headed south to meet them, but he had not been told of any such movement on the river. Union gunships are not going to escape anyone's notice; he would have known by now if one was on its way. After the Columbine's sinking, the Union was much more cautious in its use of its gunships on the river. Nonetheless, he had to allow for the possibility of river transit either by the raiders or by a gunship, positioning his artillery and some pickets along the river just in case.

Dickison has correctly figured that these sly raiders are moving quickly toward St. Augustine via land, using the river – a natural wall of water – to hinder Dickison's pursuit.

Dickison states that he was delayed ten hours by the windstorm at Palatka. He does not seem to have waited for the windstorm to fully abate based on his description of their difficult crossing. The crossing was dangerous and arduous when they decided to go for it. This 10-hour period seems to include the time during which Dickison and his men were idled, perhaps five hours, as well as the time it took to complete crossing the still choppy and dangerous river, perhaps another five hours.

When the crossing is completed, it is probably about 9-10 pm on March 11. Dickison believes he still has a chance to catch the Union invaders and seize all of what they have taken.

For the raiders and their caravan, so far, they had traveled from the Ocklawaha River bridge through the scrub 30 miles to Fort Gates, crossed the St. Johns River, and then walked another 25 miles to come to the vicinity opposite Palatka. Exhaustion is surely hitting all of them as the caravan pushes onward. They have made no stops since crossing the river and have eaten nothing either. From opposite Palatka, they still have about 25 more miles to travel to reach the safety of the St. Augustine garrison.

A road through the scrub.

Sergeant-Major James can breathe a bit easier as they move several miles north past Palatka on the opposite (eastern) shore from his pursuer. Never being certain about where Dickison was, he now knew that Dickison could only appear behind them. If Dickison had crossed the river ahead of their arrival opposite Palatka, the raiders would have been ambushed by now. There was no sign of Dickison; the caravan had a chance to make it.

Chapter Twelve

The Last Leg of the Journey

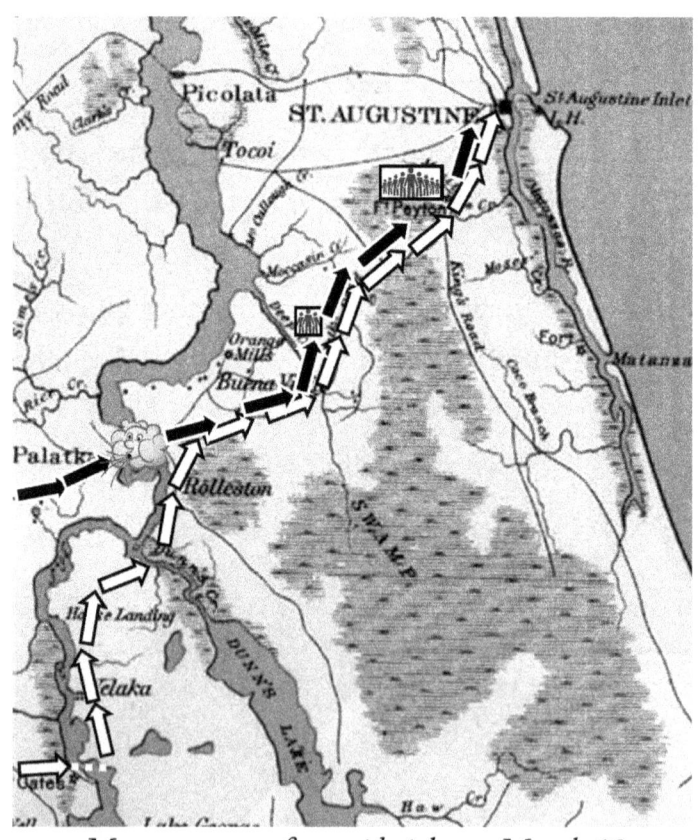

Movements to after midnight on March 12

By the time Dickison had finished crossing the St. Johns, the raider caravan was over 15 miles ahead. However, with somewhat rested horses, Dickison can close that gap quickly. And he does.

As the distance between Dickison and the raider caravan closes, the accounts of Harmon and Dickison differ.

First, Dickison's version:

> *I marched all night and at half speed and reached Fort Peaton* [or Fort Peyton], *7 miles from St. Augustine, where I overtook four negroes. We continued at fast speed toward the city and within a mile of their picket line captured twenty more, also a wagon and six ponies.*

In Sgt. Harmon's account, there is no mention of four freed slaves dropping out. On the other hand, while managing a caravan of over 120 people who have all traveled about 70 miles with no provisions and no stopping, to have a couple of the freed slaves drop out from sheer exhaustion would be no surprise. And these few people may not even be noticed by those leading the long caravan that could easily have extended over 100 yards, winding its way through the scrub of northeast Florida.

We can take Dickison's word that several freed slaves were picked up along the way, the exact number and the location where they were picked up being uncertain.

Fort Peyton appears on maps of the era located 7-8 miles from St. Augustine. Fort Peyton, an old, abandoned fort from the Seminole

Wars, was merely a landmark footnote in a wild landscape even in 1865. It no longer exists on maps today, having been absorbed by the expansive St. Augustine metro area of the modern era. The sheer lack of anything noteworthy in that area in 1865 means that Fort Peyton remains a landmark on the way to St. Augustine, but likely nothing much else.

Sergeant Harmon also mentions being seven miles from St. Augustine when freed slaves drop out, thus in the vicinity of Fort Peyton, agreeing largely with Dickison's account of a location where slaves were captured by Dickison.

Here is Sgt. Harmon's text:

> ... *by the time they* [the raiders] *had got one day's start, Dickerson's* [sic] *guerrilla cavalry were in full pursuit, and, when within seven miles of St. Augustine, the enemy overtook some of the colored people, who were unable to keep up, 19 in number.*

Where Dickison has noted the capture of four freed slaves about 7 miles from St. Augustine at Fort Peyton followed by a pursuit "within a mile" of St. Augustine when 20 more were seized, Harmon has no mention of the four but states that 19 were overtaken by Dickison's cavalry near Fort Peyton – 7 miles from St. Augustine. Harmon's number nearly matches Dickison's count, and he definitely knows where they were captured and that they were captured by Dickison.

We can give both accounts a measure of affirmation, that Dickison did pick up a few stragglers who had essentially dropped out and could not keep up, and that 7 miles from St. Augustine, around Fort Peyton, 20 more freed slaves were gathered up by Dickison.

There is no sense in either account that these 20 freed slaves had fallen away from the caravan individually or in small groups. Rather they appear to be a single cluster of folks who stopped together.

One might question Harmon's account of the reasoning for 20 freed slaves dropping out of the caravan when they were just 7 miles from St. Augustine and securing their freedom. Having gone over 70 miles, and while indeed exhausted, hungry, and likely cold as yet another mid-March night has come over them, do they drop out with just 7 more miles to go? Doubt seems warranted, although it is highly possible that 20 of them literally could not continue any further with all agreeing to drop out together.

Mitigating the reason given by Harmon's account – "unable to keep up" – is the raiders' awareness of Dickison's cavalry bearing down on the caravan. Sergeant Harmon's letter indicates clear identification of exactly who was pursuing the raiders – their nemesis Dickison – and the letter reflects awareness that it was Dickison who captured the freed slaves.

The raiders must have observed the cavalry unit approaching through the scrub in the moonlit night, and a spyglass could sight the pursuers on horseback miles away. The movement of 40-50 cavalrymen churning up a cloud of sand in this wild, flat, and desolate landscape outside St. Augustine would be noticed. On horseback, the cavalry is moving roughly three times as fast as the weary travelers of the caravan. Dickison and his men will be upon the caravan shortly.

Although the Union raiders are veteran fighters and have shown to be quite capable of defending themselves, they would not be facing a ragtag assembly of the Home Guard, but a force of combat-ready veteran Confederate regulars – perhaps double the raiders' number – plus the raiders would have all the freed slaves exposed, endangered, and ultimately a liability for engaging in a firefight. Along the way, they had lost one killed, 2-4 wounded, and three captured from the raider team, leaving them with a diminished complement of fewer than 25 of their men ready to fight at best.

Sergeant-Major James realizes that he will soon be put in the position of having to surrender, the very last thing he wants to do, made all the more troubling since they are just 7 miles from St. Augustine. He might have shared this painful insight with the whole group.

Seeing the whole enterprise imminently jeopardized, and surely grateful for the daring courage and fierce determination of their liberators, the freed slaves may have gathered to talk together, having struggled this mighty distance amid deprivation and beyond exhaustion for the sake of their precious freedom from slavery and the rigors of the plantation.

These newly freed slaves may have come up with a self-sacrificial plan: keep Dickison and his men busy and distracted for a while so that the rest of those in the caravan can make it to safety. It may have been that some of the elders in the group are willing to step up and run interference, tying up Dickison long enough to give the others a better chance of sprinting to safety and freedom. These freed slaves would also represent a prize for Dickison to claim and perhaps drop his pursuit.

The above scenario is entirely speculative. However, it would explain why these freed slaves together stopped while so very close to their destination.

It would also explain why a wagon and a group of horses were left behind with them as reflected in Dickison's account – see below. The wagon and horses would have been valued for making a run to St. Augustine. That these assets were left behind adds to the notion that this was a planned distraction to deflect the Confederate cavalry coming fast upon the caravan.

It would also explain how Dickison's cavalry failed to overtake the caravan while still 7 miles from St. Augustine.

If Dickison's cavalry was close enough to be seen in the distance in the dark, there could only have been five miles or so for the cavalry to cover – maybe 20 minutes away. There is no way the caravan is going 7 miles in 20 or even 30 minutes. However, with a substantial distraction to delay Dickison's men, the caravan will be much closer to St. Augustine and the Union picket lines, perhaps close enough to deter their dogged pursuer from chasing them further.

Why would Sgt. Harmon's account present this event simply as freed slaves being "unable to keep up"?

First, it would be accurate in a sense, reflecting the desperately worn-out state of all the travelers as they come close to their destination. If these were the elders among the freed slaves, the physical toll on them would be particularly acute and the reasoning would reflect an accurate statement of their sheer exhaustion.

Further, this may have been the reasoning presented, telling the raiders that they were simply too exhausted to continue and that they were dropping out.

Second and most importantly, it detracts from Harmon's narrative of Black soldiers succeeding with their strategy, their leadership, their courage, their determination, and their skill. Shifting the credit for the mission's success at the end of the story to these 20 freed slaves who make a sacrificial choice to ensure the raiders' success spoils the overall narrative. Harmon may have shaped the descriptions to fit the narrative, a common practice for anyone telling a story.

Here is Dickison's statement again:

> *We continued at fast speed toward the city and within a mile of their picket line captured twenty more, also a wagon and six ponies.*

Dickison may have erred (deliberately?) in his recollection again, conflating the capture of 19 freed slaves, plus a wagon and six (loose?) horses, together with a (resumed) pursuit to within a mile of the Union picket line before calling off the chase. It is certainly more dramatic to seize escaped slaves only a mile from their destination of sanctuary, and we know Dickison likes to be dramatic, particularly in a rather sour report to his commanding officer in Tallahassee. However, it is unlikely to have happened in the way he described.

Both Harmon's and Dickison's accounts pivot on what happened 7 miles from St. Augustine near Fort Peyton. The nod should go to Harmon who seems certain of the number dropping out, where it occurred, when it occurred, and finally that they were indeed captured by Dickison. Dickison's report jumps around at the end as he tries to salvage his own narrative that would benefit from a bit of drama, a tall order given the failures of his pursuit.

Using the darkness around midnight to full advantage, imagine Confederate cavalrymen riding up to confront stirred-up horses and a wagon splayed across the narrow path while a distraught group of 20 desperate slaves makes a terrific scene, some pleading for mercy and assistance while others dash off into the scrub, perhaps riding the horses, inviting the distracted pursuit of the cavalrymen. The contrived chaos would likely sow plenty of confusion even among the veterans in Dickison's command.

Distracting Dickison and his men, interrupting his headlong pursuit by bringing it to a complete stop, forcing him to sort out the confusion, command his men to restore order, and determine what to do next was buying precious minutes while the rest of the caravan sped ahead.

This delaying tactic may have proven to be just enough, both in consuming time as well as presenting some consolation prize for

Dickison to claim. Unknowingly, Dickison had taken the raiders' bait again and followed their script.

By comparing Dickison's and Harmon's accounts, we can imagine how very close they were.

Dickison and his men finally get a grip on the situation near Fort Peyton, leaving some of his men to guard the captured slaves and deal with the horses and wagon as the rest of his unit resumes their pursuit.

As the cavalry closes the distance to the caravan, Dickison also realizes that they are coming very close to the Union picket lines. They do not want to deal with Union pickets plus likely reinforcements from the garrison. Dickison's men were also exhausted, and their horses had done plenty of hard duty to get them this far.

Union picket camp near Bull Run in Virginia

The pickets were too close, and the caravan had gone too far at this point. Dickison decides that it is wiser to accept what they have gotten rather than pursue a risky course of committing his weary men to combat with a superior Union force. Such an extended endeavor is unlikely to achieve the recovery of much more (if any) of what the raiders had taken while also exposing Dickison and his men to needless casualties. And yes, even Dickison knows that this war is nearly over; it simply isn't worth it.

Dickison and his men turn back, collect what they had retaken, and call it a relative success. It wasn't. Dickison had been beaten by these raiders and he knew it. Worst of all in Dickison's mind, it was done by Black soldiers.

Old gates of St. Augustine, one looking into town, the other looking out. Note the absence of any trees on the right. A perimeter cleared of anything that might provide cover seems to have been practiced around the town.

We can imagine the shock of the Union pickets outside St. Augustine as the noise of a wagon, horses, mules, and people – over 100 people – emanates from the scrub, a large body of silhouettes moving quickly through the darkness toward the town. Brought to alert, we imagine the pickets challenging the mass of travelers. The raider soldiers would call back with their unit identification and from whence they have come while also warning of the threat to their rear from the pursuit of Dickison's Confederate cavalry.

Then, we might expect a more general alarm to sound as soldiers snatch their muskets and gear as they awaken from slumber and come to the road, perhaps a few in their underwear since this is all happening after midnight on March 12.

Wisely, Dickison has dropped his pursuit, knowing that nothing worthwhile could likely be gained and serious trouble could await him in a short distance if he continued.

Most of the caravan finally reached its sanctuary behind friendly lines. Mission accomplished.

Chapter Thirteen

Commentaries After the Raid

In the end, the raider mission was a stunning success, enabling Sgt. Harmon to note proudly and triumphantly:

> *The remainder of the party reached St. Augustine on the 12th inst., in safety with the wounded, 4 prisoners, 74 liberated slaves, 1 wagon, 5 horses, and 9 mules, having travelled over 200 miles of the enemy's country, doing without food for 3 days, and 100 miles of our own country, in five days and nights, reaching Jacksonville last evening, the 19th inst., with all their booty.*

The observant will note that Sgt. Harmon's net number for the liberated slaves and the reported number is 74. He lists 95 coming away from the Marshall Plantation, 19 dropping out around Fort Peyton, and 74 arriving in St. Augustine. His count is off by two freed slaves.

Later Union accounts will report 71 freed slaves. This seems to be the best number since Dickison consistently reports seizing 24 slaves altogether. A count of 71 freed slaves is solid, and that is quite a large number for this operation.

For this intrepid team of Black soldiers and scouts, their primary purpose was to employ a daring strategy of their own design, lead their own operation, and succeed in beating the Union's nemesis Capt. J. J. Dickison, something the White commanding officers had never been able to accomplish, except for Dickison's ill-fated attack on Gainesville in February 1864 prior to Olustee. The raiders succeeded on all counts.

Naysayers may point to the windstorm delay for Dickison in Palatka, asserting that the raider caravan would have been easily caught and captured if the weather had been more normal. If.... If.... If.... Bold complex plans with multiple contingencies are laden with risks, opportunities, and misfortunes. There are many "ifs" that could have impacted the outcome for both sides. In the end, the "ifs" don't matter; it didn't happen that way.

The mission was not flawless. The abortive venture onto the Holly Plantation stands out as the most tragic mistake. The huge amount of territory for the raiders and their caravan to cover may not have been accurately understood. The difference between the caravan walking and the cavalry on horseback may also not have been accurately assessed. The scouts, who were likely quite influential in their assessments of how things would unfold, may not have considered the slower pace of a caravan versus moving a small number of escaped slaves to St. Augustine. Again, there are contingencies in any plan which can determine the outcome one way or another.

Indeed, Dickison had never been beaten like this since his terrible debacle in Gainesville over a year prior as far as we can tell. He routinely had the upper hand, enjoying and exploiting native privileges of terrain familiarity and information-gathering with regular, resounding success against Union forces. Not this time. Despite the setback, he will try to wring out some good news from all this:

> *We continued at fast speed toward the city and within a mile of their picket line captured twenty more, also a wagon and six ponies. Three of these ponies have since been claimed by citizens and delivered to them. The enemy, on hearing we were in pursuit of them, left wagons, mules and provisions at the river, where they had crossed near Fort Gates.*

Dickison spins the story in his report. Reaching back in time, Dickison suggests that the raiders were struck with fear at the Home Guard's approach back at Fort Gates. (Dickison slips into "we" here to describe the efforts of the Home Guard who shared in pursuing the raiders. His messages from the Home Guard must have also mentioned the militia's capture of the broken-down wagon in the scrub and the horses (mules? provisions?) left behind by the raiders at Fort Gates.)

This description is rather amusing, knowing what we have learned from Sgt. Harmon's account; one wagon broke down in the scrub on the trek to Fort Gates and only nine horses were left behind at the flatboat landing's west side. Provisions? Whatever it was, it was left behind, not exactly taken from the raiders, while most of what was truly valuable, the slaves, escaped to freedom.

Dickison's description of the destruction of the Marshall Plantation only noted:

> *The raiding party on reaching her* [the widow, Mrs. Marshall's] *plantation destroyed 200 hogsheads of sugar.*

This glaringly omits mention of the complete destruction of the entire plantation's working operation and the loss of extensive stores of varied kinds that were stocked for processing or delivery. The plantation operation never recovered or resumed; its destruction was complete and final. Yet even Dickison had to admit that a large quantity of valuable stores was destroyed.

Also never mentioned is the full number of slaves that had been freed from the plantation, leaving later Confederacy apologists to assert that the freed slaves that Dickison had seized constituted all the slaves of the plantation. Wrong.

Dickison will continue, planting a seed that would lead others to claim wrongly that the raiders were confronted at the plantation:

> *Some of our militia* [not his cavalry unit] *met them* [the raiders]*, and in an engagement two of our men were killed.*

Dickison seems to be referring to the night battle in the scrub before the caravan reached the Fort Gates landing. Dickison's claim of two Home Guard fatalities contrasts with Sgt. Harmon's claim of 20 dead. Neither is probably correct, but the reality is likely closer to the four fatalities noted by *Ocali Country*. No mention is made that the militia – the Home Guard – was quickly scattered. No mention is made of how many in the Home Guard deployed.

Overall, the Dickison report seems intended to minimize the costs inflicted by the Union raiders while indicating that an effective defense against them had occurred None of that would be true – the costs were high and the defense against them was quite ineffective.

Engaging in the kind of "what if" conjecture noted earlier, Dickison whines rather defensively, aware that, while he may be putting up a good front for his commander and any other readers, he knew that he had been beaten:

> *[If] information reached me earlier they would have been overtaken with their rich spoils before reaching the river.*

"Before reaching the river"? This only makes sense if Dickison is now referring to the San Sebastian River which is the natural boundary on the western side of St. Augustine, a river that the raider caravan needed to cross to get to the safety of the town.

At least Dickison admits that the raiders absconded with "rich spoils," the closest he comes to acknowledging their success.

It is interesting to note that he faults the timing of information yet makes no mention of the misfortune of the weather preventing his rapid river crossing at Palatka. While Dickison is the only one to cite the windy weather at Palatka, he is also the only one in the story who would have been affected by the wind.

Among Dickison's closing remarks are the following, typical of his reports:

> *The march was truly a hard one. We marched four days and nights with but little forage or provisions. My men were resolved, and showed a determination to pursue the enemy to the very gates of the city… All praise is due these noble, gallant men for their unflinching spirit and resignation, having endured every hardship without a murmur.*

One final note about Dickison's account is the absence of any mention that the raiders were Black soldiers. Surely, he knew. For Dickison, the most damning feature of this highly successful Union engagement, using such a clever and audacious strategy to fool him into following their script, would be that it was all done by Black soldiers. Dickison readily acknowledges the presence of "negro soldiers" in other accounts within his historical volume about the Confederate military in Florida during the Civil War since Black soldiers were common among Union operations in Florida. Here he remains conspicuously silent about his enemy, the one enemy unit to beat him.

Indeed, the raiders took a page out of the guerilla warfare playbook that Dickison had employed so successfully. The raiders exploited local knowledge in the intelligence gathering that discovered the flatboat near Fort Gates, knowledge of a large number of slaves at Marshall's, and used the geography of the wide river together with a secretive strike further south than any previous operations to make the raid a success.

Dickison unwittingly followed the raiders' script on two occasions. The first was his futile scramble to get to Marion County while the raiders made their escape across the St. Johns River. The second was his unit's apparent distraction with the contrived delaying tactic by Fort Peyton. Both strategic ploys enabled the mission's success.

One final note about the freed slaves. Mentioned earlier for its incorrect characterization of the raid, the letter of Mrs. H. B. Greely to her associate "Br. Whipple" does provide her first-hand report on a relevant factor. Mrs. Greely seems to have headed the work of the American Missionary Society in St. Augustine, dealing with a host of people – Whites and slaves – and their variety of needs in this Union garrison town which provided sanctuary for many who were left with nowhere else to go to survive. This included liberated slaves. Here is the first paragraph of her letter (the year is misstated; it should be 1865):

St. Augustine, Fla.

Mar. 18th-64

Dear Br. Whipple

... Twenty five Colored men from Col. Tighlman's [sic – Col. Benjamin Chew Tilghman, commander of the 3rd USCT] *regiment stationed in Jacksonville went out on raid last week in this state, and brought into this City, Seventy contrabands.* **They were the most destitute objects I ever saw. Many of them almost entirely naked.** *The teachers of the F. R. Ass.* [Freedmen's Relief Association] *having some money resulting from the sale of books & c. and other means which we could raise, have been very busy this week in making up clothes for the women and children, and we shall soon have them in comfortable condition to remain here, or go elsewhere as Government may see fit to dispose of them. The raiders also brought in some fine horses and mules.* [Emphasis added.]

Mrs. Greely's description of the freed slaves from the Marshall Plantation reveals starkly how desperate the conditions in Florida had become as the war came to a close. Many deprivations came to those in war zones as well as to those far beyond them. Marion County was far from military action throughout the war, yet its White inhabitants reported dire conditions in the last months of the war. If conditions were difficult for the White masters and overseers, it is painful to imagine the wretched treatment of slaves. Mrs. Greely's description of the freed slaves' condition is likely no exaggeration. Surely, she had encountered plenty of struggling, suffering, and damaged people. The awful state of these freed slaves shocked even her. (The full text of her letter is included in the Appendix.)

Let us close with Sergeant Harmon's closing to his letter:

> *This expedition reflects great credit on Sergt. Major James, for the masterly manner in which it was commanded, and gives further proof, that a colored man with proper training can command among his fellows and succeed where others have failed. And a great deal is due to the men for their good behavior, and steadiness, and obedience, and if it were not for occupying too much of your space, which I fear I have done already, I would give their names, but that at some other time.*
>
> *I am still an ardent lover of my race, and a soldier.*
>
> *H.S.H.,*
>
> *Sergeant Co. B., 3d U.S.C.T.*

Battery McCrea, Jacksonville, Fla.,

April 3d, 1865.

Chapter Fourteen

Conclusion

The success of these raiders did not go unnoticed. Here is what was reported as found in Union records recounted in *Freedom by the Sword* (p. 87; p. 107 PDF). The full text of the General Orders from the office of Major General Gillmore is provided in the Appendix:

> *"I think that this expedition, planned and executed by colored Soldiers and civilians, reflects great credit upon the parties engaged in it,"* the regiment's commanding officer wrote to General Gillmore, *"and I respectfully suggest that some public recognition of it, would have a good effect upon the troops."* The letter went to department headquarters, where, on 20 April, Gillmore praised the raid in general orders: *"This expedition, planned and executed by colored men under the command of a colored non-commissioned officer, reflects great credit upon the brave participants and their leader. The major-general commanding thanks these courageous soldiers and scouts,*

and holds up their conduct to their comrades in arms as an example worthy of emulation."

The final report[1] of the raid acknowledged two deaths and four wounded among the raiders; no mention is made of the scouts who were captured at the Holly Plantation. Of course, the scouts were non-military or "civilians" as Sgt. Harmon labels them and would not necessarily be included among the military participants or casualties.

Gen. Gillmore's Hilton Head, SC HQ

With no other shooting action to produce a fatality after the encounter with the Home Guard in the scrub, we can assume that one of those wounded in that gunfight died from their wounds. The other fatality was disclosed by Sgt. Harmon, the death of Sgt. Joel Benn during the failed Holly Plantation foray.

Again, the stated purpose of this outside-the-box, dangerous raid so late in the war – among the last organized Union military actions in Florida – is not directly disclosed.

---

1. Gen. Gillmore's general orders were dated April 12, 1865. It reflects the deaths of Sgt. Joel Benn in action at the Holly Plantation and most likely Pvt. Francis Fisher who "died of wounds received." His death is recorded as April 4, 1865. Sgt. Harmon's letter is dated April 3, 1865.

The White officers acknowledge the raid was "planned and executed by colored men under the command of a colored non-commissioned officer," a reminder of the uniqueness of such an operation, its design, and its leadership, but also reflective of the evolved attitudes toward Black soldiers and their capabilities during the course of the war. Having previously witnessed the excellent deportment of Black soldiers in combat, agreeing to let a motivated group of Black soldiers carry out their own plan for a mission may have seemed a natural progression to the Union command in Jacksonville.

These were key elements that the raiders were determined to prove; their ability to plan, organize, lead, execute, and succeed in a most difficult and challenging mission. Certainly, the raiders took justifiable pride in the acknowledgment of these goals by the Major General commanding the Department of the South, US Army.

The less explicit piece had to do with Dickison, beating Dickison in their raiding mission, something the White leaders had been unable to accomplish. This team of Black soldiers and scouts wanted to show their mettle in a particular mission explicitly designed to beat Dickison. Letter writer Sergeant Harmon, being active military, cannot directly crow about the achievement without stepping on his White commanding officers' toes. He and his unit could end up with some ugly assignments as a result. A bit cheeky, he tips us off when he writes, "and succeed where others have failed." Knowing the context and a major object of the mission – beating Dickison – we can understand what Harmon meant when he wrote that.

Medals for a particular service were a new thing. The first Medals of Honor were issued in 1863 during the war by President Lincoln.

Most medals for the Civil War were struck long after the war had ended. Therefore, we should not expect any medal to be promoted for the raiders. The mention in the Dept. of the South, Union Army commander Major General Gillmore's general orders is as good as you can get.

Just three days before the praise of the raid was issued by General Gillmore, Confederate General Lee had surrendered to Union General Grant at Appomattox on April 9, and just two days later on April 14, President Lincoln would be assassinated. The daunting task of rebuilding the nation after the catastrophic civil war became utter chaos as the steadfast wartime leader, Lincoln, was murdered and a man wholly incompetent, unprepared, and bizarrely erratic even when sober, Andrew Johnson, assumed the nation's leadership. The strange, successful raid of the Marshall Plantation by Black Union soldiers led by a Black non-commissioned officer beating the Confederate nemesis Dickison is eclipsed by the celebrated end of the war and then the shattering event of Lincoln's assassination, all amid the first weeks of April 1865.

The Marshall Plantation raid would escape the notice of almost everyone as the nation moved ahead. Florida was almost irrelevant militarily in the Civil War to begin with, and this last raid, stunning in its details, its context, and its uniqueness, particularly for Black Union soldiers, seemed like an odd end note and was largely forgotten.

Hopefully, the story will continue to be re-told, highlighting the bold and thoughtful strategy, the bravery and determination of the men, and the importance of the achievement for these Black soldiers in seeking to gain recognition, standing, and equity with their White peers. It is a struggle that continues for all Black people today.

For the soldiers involved with this mission, its success marked a milestone in their journey that had started two years before. After their training, they would see a variety of action in South Carolina, briefly in Georgia, and the prior year in northeast Florida. From the most mundane, boring, and exhausting tasks to adrenaline-driven, dangerous combat confrontations with the enemy, these men had earned the respect of their fellows and their superiors.

Let us also remember that nearly half of the raider team – the six men from the 34th USCT (former 2nd SCVI) and the seven "scouts" or guerilla fighters – had likely all been enslaved previously and were working through their first years of freedom from a White master.

A slave takes orders from a White master or else. The slave may know better ways, have creative thinking, and have the capability of organizing and leading, but any kind of assertiveness is likely to lead to trouble and punishment, stepping out of place in a rigid social hierarchy.

The military can also be a rigid social hierarchy and seemed stacked against Black men, reserving power, leadership authority, and privilege for White officers, excluding Black men regardless of their capabilities. Stepping out of place in the military could also have severe consequences.

The conditions were very different between the two settings, but the socio-cultural structures of the plantation and the military must have been noticeably reflective of each other for these former slaves.

This Black-planned, Black-led, Black-executed mission was a clear departure from that socio-cultural structure, indicating that there was indeed a different path forward, a structure that could and should include Black soldiers in the commissioned officer hierarchy that had been reserved for Whites only. That different path was quite simply equality. These former slaves were ready to risk it all once again in the effort to seal their liberation with actions that defied the socio-cultural structure and established equal footing with Whites.

They had all encountered leaders like General Seymour who regarded Black soldiers as expendable fodder in order to enable action by White soldiers who were regarded as 'real' soldiers. They had also encountered thoughtful and wise leaders like abolitionist minister Colonel Thomas Wentworth Higginson who led the 1st South Carolina Volunteer Infantry (later 33rd USCT), and who was a mentor to Emily Dickinson, writing: "We, their officers, did not go there to teach lessons, but to receive them. There were more than a hundred men in the ranks who had voluntarily met more dangers in their escape from slavery than any of my young captains had incurred in all their lives."

As enlightened, respectful, and encouraging as some White leaders might be, there were no Black leaders. When strategic planning of operations occurred, there were no Black men hovering over the maps engaging the conversation. Black men were the manpower behind those plans developed by White men, and Black men were the targets in the front-line combat being orchestrated by White men. The Black men serving to advance the Union effort to end the White supremacist-slavery regime of the secessionist states were fully aware of their carefully circumscribed role which would eventually invite them as participants first, then warriors, but with a strict ceiling on planning and leadership roles.

Martin Robison Delaney, a civil rights activist, was commissioned as a major in February 1865, becoming the first and highest ranking Black field officer in the U.S. Army

The unpredictability of the battlefield would often put Black soldiers in ad hoc command roles which they would normally be denied. These Black soldiers could rise to the critical occasion and demonstrate their abilities. After the loss of White commanding officers in harsh combat situations, likely amid a dangerous threat to the whole unit, Black soldiers were repeatedly shown capable of rallying their men, defining a proper course of action, and leading their unit through the situation.

Surely these instances were known and influential as Black soldiers began to be commissioned as officers as the Civil War was coming to a close. Perhaps 100 Black soldiers, excluding chaplains, were commissioned as officers during the war – all within the last few months of the war – out of roughly 200,000 Black soldiers who served in the Union Army.

In the context of the journey of the Black soldier, the Marshall Plantation raid is one of those milestones in a greater movement forward. While the meaningless raid occurring in a meaningless operational theater and at an irrelevant time left it unnoticed, sidelined with sufficient reason from a practical viewpoint, it nevertheless is a milestone story of Black soldiery, showing the ambition of Black soldiers to claim roles in planning and leadership, to demonstrate strategic thinking and accept visionary risk, to move boldly and courageously in ways others had not considered, and to "succeed where others failed."

On one hand, this was an obscure and seemingly irrelevant mission in the scope of the war, but on the other hand, a noteworthy achievement in the journey of the Black soldier. From the example of men like these, Black soldiers would slowly progress. Much too slowly, of course.

The role of Black people in the US military continued to be subject to various forms of circumspection based on racial biases for more than 100 years. It would be nearly 150 years before there was a Black commander-in-chief, a Black Secretary of Defense, and a Black Chairman of the Joint Chiefs of Staff.

Chair of the Joint Chiefs Gen. Colin Powell; President Barack Obama; Secretary of Defense Lloyd Austin

Marking the way forward was a small raiding team of mostly Black men in Florida who had the audacity to plan, organize, lead, and execute a raid 100 miles behind enemy lines, directly challenging a noteworthy, accomplished foe, and succeeding. Only two years from the start of the experiment in Black recruitment, these men showed that they were capable of doing it all on their own, that Black soldiers simply needed the door opened to the opportunity of those leadership roles. The Black soldier was now ready, having proven it to themselves and to any who may have doubted.

Salute the men of the Marshall Raid and celebrate their stunning victory.

Chapter Fifteen

Postscript

Pre-Civil War Henry James

A friend who does extensive genealogy research, Kim, has been a fan of the book and was curious to know more about Sgt. Major Henry James.

I, too, would love more information about him. Sergeant Major is a regimental position, making a Sergeant Major in a regiment like the 3rd US Colored Troops the highest-ranking Black soldier. Having been appointed to his position roughly 10 days after his enlistment, there must have been certain qualities that made him the choice. All we know from his enlistment document is that his occupation was listed as "laborer," a term recorded for most Black enlistees. So, what were those qualities that made Henry James outstanding?

I knew that a Sergeant Major's position demanded extensive bureaucratic duties, requiring numerous reports on all kinds of subjects as well as ensuring the effective drilling of the men in the 10 + companies of soldiers in the regiment. The Sergeant Major had to be literate, organized, and disciplined.

I admit that I had not vigorously pursued background info on Henry James. I had my reasons. He was not a commissioned officer. He was not auspicious in any real way. His name was common and could be seen as two first or two last names, causing searches to return a huge number of results with nothing coming close. Finally, any records would be roughly 150 years old. This pursuit didn't seem to be a good use of research time.

Kim states that in a matter of minutes - MINUTES! - she had struck gold. I was caught between historian humiliation and historian hallelujah. While I hang my head in shame, let's pursue the hallelujah.

The link on my website – bruceseaman.com/updates – is to a 12-year-old article that recounts an interview with Henry James in 1891, 4 years before he died in 1895, appearing in the Chester County (PA) Daily Local News, presumably in 1891.

It is a long and fascinating article that makes no mention of his leadership of the March 1865 raid on the Marshall Plantation. As has so often been the case in this research, a whole story of its own unfolds.

Let me summarize.

At the 1860 census, Henry James lived in Ercildoun, PA, a hamlet in Chester County outside Philadelphia founded by Quakers and an early center of the abolitionist movement. In 1860, he was hired as a "body servant" to Lt. Chambliss of the 5th US Cavalry. The unit would be stationed at Camp Cooper, 150 miles west of Fort Worth, TX, a place known for its hellish heat and brutal conditions.

In early 1861, Texas seceded from the Union.

[quoting the Main Line Today article]

> *By Confederate logic, all blacks were someone's property. So, if James didn't belong to any of the men, he must be federal property. Chambliss realized that his employee would likely be seized and sold.*

> *"When Lt. Chambliss found that a surrender was inevitable, he told me that there was only one chance for me to escape—and that was to become his slave," James later related. "He could have sold me after I had once voluntarily become a slave, but I trusted him and he was true to me."*

> *Events unfolded as Chambliss had expected.*

> *"Every Negro found in camp was asked, 'Who do you belong to?' I promptly proclaimed myself as belonging to Lt. Chambliss. The lieutenant acknowledged the ownership and I was duly inventoried among the personal belongings of the lieutenant," James recalled.*

With the Union garrison ordered out of the state, Chambliss and James journeyed back to Pennsylvania. James had to play his role effectively.

[again quoting the Main Line Today article]

> "I had a tolerably good education," [James] related. "But the lieutenant cautioned me not to display my knowledge. 'Don't read a paper,' said he, 'and be as stupid as possible.'"

Returning to Union states in April 1862, the trek was apparently uneventful. James worked for a year as a wagon driver before enlisting in the 3rd USCT in mid-1863.

[quoting the Main Line Today article again]

> "My close observance of army affairs while serving as body servant for Lt. Chambliss made me familiar with the drill," he explained. "So I was proficient in the duties of a soldier."

Now the appointment of Henry James as Sergeant Major of the 3rd USCT makes perfect sense.

Not much is known about Sergeant-Major James after the war. Journalist Rick Allen's research in his February 2016 article is a solid description:

The war over, the black units were disbanded; the 3rd U.S.C.I. on Oct. 31. James joined the newly formed 9th Cavalry and served in Texas as a "Buffalo Soldier."

["Buffalo soldiers" were apparently given the name by First Peoples-Native Americans because the wiry hair of these Black men was reminiscent of the hair on buffalos.]

He never regained the fame of the Marshall raid. His squad was caught in a hailstorm and he developed a severe infection, according to an account compiled by the 3rd U.S.C.T. Re-enactors. In 1869 he was discharged for medical reasons and he returned to Pennsylvania. In 1887 he was awarded a disability pension, and at some point he married. Rachel G. James is listed in his official pension records as his widow.

He died in 1895 in Carlisle, Pennsylvania, and is buried in a nearly lost grave at the Mount Zion A.M.E. Church cemetery in Atglen, near where he grew up.

Sgt-Major James headstone

Captain J. J. Dickison

Dickison would be given a long-delayed promotion to Colonel just before the Confederacy dissolved entirely in May 1865.

Dickison was an ardent promoter and defender of the Confederacy. He wholeheartedly embraced the Confederacy's *raison d'etre* with its White supremacist ideology. This is evident in his history writing. Following his second encounter in Gainesville when he routed Union troops and captured nearly 200, he noted this:

> *Our largest and most productive interest—sea island cotton—and the immense supplies of corn and forage, made it of the highest importance that this wide extent of country should be closely watched and the advances of the enemy checked,* **preventing widespread desolation and the carrying off of the slaves, who were the only able-bodied tillers of the soil, and better fitted for field work than the white man.** [Emphasis added]

When Dickison and his men scuttled the USS Columbine in May 1864, it is believed that Dickison had submerged at least one of its large lifeboats for future use. It is believed that this lifeboat was raised a year later and used to smuggle Confederate General and Secretary of War John C. Breckinridge to Cuba, and perhaps other high-ranking Confederate officials. Many of the Confederacy's cabinet members passed through Marion County, some staying for an extended period – enjoying the weather? – before being secreted out of the country to asylum.

John C. Breckinridge

Here is how one biographer recorded Dickison's post-war years:

> *Dickison was opposed to Reconstruction and as a segregationist got involved with the Ku Klux Klan* [and was said to have been a key figure in a Klan plot to smuggle a large store of rifles and ammunition into Jacksonville to foment an insurrection]. *His character is described as volatile, irascible, unstable; people admired him for what he did during the war, but they preferred keeping their distance.*

> *Having lost his wealth (four years of war ruined his plantation due to neglect), he had a number of occupations to earn money. He for instance became a farmer in Quincy*

while working in an asylum in Chattahoochee, he twice served in the state legislation [sic] *and in the early 1890s as treasurer of Marion County. After Reconstruction, in 1877, Governor George Drew appointed Dickison Adjutant-General of Florida's military affairs in which position he served for four years.*

Aside from financial struggles, the decades after the war were not kind to Dickison on a more personal level as his family broke apart. Robert Ling died in 1867 (aged only 16) of unexplained causes. His daughter Mary Lucille moved to Baltimore after marrying and they lost contact. His wife became seriously ill, though she eventually recovered. In 1881, Dickison took her to White Springs for recuperation when news arrived that John Jackson Dickison, Jr. had been accused of murder and had been lynched by friends of the victim (he was only 28 or 29 years old). [Also, Dickison's 18-year-old son Charles, a sergeant in his father's Company H, was killed in action by Union forces in 1864.]

When in 1888 the United Confederate Veterans organized the Florida division, Dickison was elected as the state commander. He was reelected five times before retiring. He held the state military title of Major-General. [This is the rank/title on his headstone.]

In 1890, his wife published the book "Dickison and His Men", describing in great detail the various exploits of Company H of the 2nd Florida Cavalry. Because of this, many wondered whether Dickison had actually ghost-written the book. [His wife's biography also reproduces his report to his commanding officer about the Marshall Plantation raid. It is almost verbatim, but the numbers of slaves in this earlier biography stated 3 freed slaves were picked up instead of 4 (he claims at Fort Peyton), and that 21 freed slaves were later picked up instead of 20 (he claims within a mile of St. Augustine)].

In 1899, Dickison wrote the Florida volume of the Confederate Military History. It is still regarded as an important reference for Civil War actions in Florida because Dickison used documents that are no longer available.

Dickison died on August 23, 1902 at the age of 86. Cause of death was typhoid fever although his health had been declining for many years. On August 26, many shops in Jacksonville closed, and thousands lined the streets to watch the funeral procession. Dressed in his Confederate uniform, Dickison was laid to rest in Evergreen Cemetery in Jacksonville, Florida.

[unidentified author] – see link –

https://civilwartalk.com/threads/the-swamp-fox-of-the-confederacy-%E2%80%93-john-jackson-dickison.170566/

The Jacksonville Mutiny

The struggle for Black soldiers to be given the respect they had earned and deserved was not settled by their conduct during the war. Even the 3rd USCT would see a ghastly outcome to a relatively trivial post-war incident.

This text is from the excellent source *Thunder on the River: The Civil War in Northeast Florida* by Prof. Daniel L. Schafer who draws information from another excellent source by Prof. Joseph T. Glathaar, *Forged in Battle: The Civil War Alliance of Black Soldiers and White Officers*, and additional details have been added in brackets drawn from the highly detailed account of John F. Fannin, *The Jacksonville Mutiny of 1865*, published in The Florida Historical Quarterly:

> The 3rd U.S.C.T. was stationed in [Jacksonville] from February 5 until October 31, 1865. A white officer in the 3rd U.S.C.T., Lieutenant C. W. Brown, kept a diary during the months he resided in Jacksonville, recording incidents of insubordination, fistfights, and violent confronta-

tions. While in Jacksonville, Brown was involved in an incident that [Prof.] Glatthaar calls "the most shocking mutiny" of the entire war. The 3rd U.S.C.T. was ordered to muster out of Union service in Jacksonville and depart for Philadelphia by October 31, 1865. On October 29, after several days of confinement in their tents while a late-season hurricane lashed northeast Florida, the black soldiers emerged to discover one of their comrades tied by his thumbs to a scaffold on the parade ground for stealing a jar of molasses from the field kitchen.

... [T]hey had been subjected to numerous types of cruel discipline in the "old-army way" during more than two years of military duty. [The customary manner of administering this punishment required that an officer strip the prisoner to the waist and tie him by the thumbs to an overhead support - his toes barely touching the ground, his quivering calves quickly turned to jelly, as his thumbs strained from the sockets.] Brown's diary documents several such punishments of soldiers in the 3rd U.S.C.T.

Witnessing this punishment yet again, only two days before mustering out, sparked a furious protest. Thirty men started toward the platform shouting and gesticulating, some carrying weapons. The white officers stood their ground until the protesters were ten to fifteen feet away, when Lieutenant Colonel John L. Brower fired three shots into the crowd [of soldiers, unarmed]. Some enlisted men [then went back to their tents to get their rifles and] shot back, and wounds were inflicted on both sides. Fifteen soldiers were arrested and confined at the Bay Street stockade. Legal officer Alva A. Knight, a graduate of Amherst College, captain of Company B, 34th U.S.C.T., and a future lawyer, judge, and state senator in postwar Florida, filed charges of mutiny in a war zone against the men, which carried the possible sentence of death by firing squad.

Only two days later, the court-martial trials commenced aboard a ship ... the St. Marys off the town of Fernandina. [Lt. Col. Brower hedged his testimony, claiming that he had fired warning shots which was readily disproven by Private Joseph Green who had been wounded by two of Brower's shots. Brower would not receive any discipline, nor would Lt. Graybill who began the incident by ordering the thumb-hanging of the thief of the jar of molasses and had also fired in the crowd of soldiers with his revolver.] Thirteen convictions were returned, including six death sentences. The condemned mutineers were executed on December 1, 1865 [on the beach at Fort Clinch]. What had begun as a petty theft of a jar of molasses ended

Hanging a prisoner by the thumbs

only thirty-four days later in [six] death[s] by firing squad.

There are plenty of other accounts of less lethal incidents involving Black occupation soldiers facing stiff and often violent White resistance to their presence, including a detachment of the 34th USCT deployed to Lake City in December 1865.

The Lake City incident resulted in the Provisional Governor ordering the Black soldiers to return to Jacksonville. Captain James Montgomery, the son of the 34th USCT's first commanding officer Col. James Montgomery of "Jayhawker" notoriety, defended his men, noting that drunken White men acted violently against the Black soldiers who were unarmed in compliance with their orders. The situation became explosive as the White civilians armed with guns and knives threatened the Black soldiers. Captain Montgomery sent an emergency company of Black soldiers with loaded muskets into the square to quell the hate-filled fever of the mob successfully. Many in the mob were former Confederate soldiers who promised that when the troops depart, they would "clean the deserters and niggers out."

Despite having proven their military prowess and courage during wartime, peacetime brought out both old and new racist attitudes.

From the defeated Confederacy-supporting Floridians for whom White supremacy had been the traditional standard, the specter of large numbers of Black soldiers freely moving about and engaging their communities was profoundly disturbing.

For White Union officers, dealing with large numbers of freed Black slaves who were destitute and dependent, plus large numbers of Black soldiers who felt they had earned a certain status of respect and recognition, racist beliefs surfaced that refused to accept parity between White people and Black people.

Both the White Floridians and the Union officers felt overwhelmed and outnumbered by the large presence of Black people both in uniform and former slaves, discovering uncomfortably how the old standards had been cast off (defeated?), producing reactions of anger and even violence.

In September 1865, there was the court martial of Second Lieutenant Henry K. Cady of the 34th USCT, a White officer, who had developed a pattern of threatening Black non-commissioned officers and soldiers. According to the records of the court-martial, Cady had said:

> *The damned nigger did not belong to the human race, and the form of his head showed it to be a fact, and that it could be proved by the Bible;" and that when asked by Captain James Montgomery "what he came into a colored regiment for," did answer, "that he came into the regiment for the benefit of his pocket, as every other man did.* (Fannin, The Jacksonville Mutiny, p. 385)

Cady was discharged, not punished.

Captain H. M. Jordan, a White officer also of the 34th USCT was tried around the same time in a court martial for sexual misconduct "with a certain colored woman," but whose guilty verdict was overturned upon review for a technicality, causing him to be restored to duty to the utter dismay of disgusted Black soldiers for whom the hypocrisy, injustice, and racial bias was evident. A Black soldier could be hung from his thumbs or forced to 'ride the rail' for minor offenses, but a White officer would receive no penalty for their gross misconduct.

Finally, there is another letter written by Sergeant Harmon whose letter to *The Christian Recorder* provided us with the thorough Union account of the Marshall Plantation raid. Undated but published in *The Christian Recorder* on October 21, 1865, Sgt. Harmon responds to another published letter that articulated the harsh punishments experienced by Black soldiers. Harmon affirms the statements made in that letter and adds his own testimony. The full text is provided in the Appendix.

It should be noted that Lieutenant Colonel John L. Brower of the shooting incident during the Jacksonville Mutiny noted above was only 22 years old when appointed to this high rank and placed in charge of the 3rd USCT detachment that occupied Gainesville in the summer of 1865. One wonders if Sgt. Harmon is reflecting upon the leadership style of Lt. Col. Brower in his October 1865 letter.

An excerpt from Sgt. Harmon's letter, published October 21, 1865, detailed the anger and resentment of the unequal and unfair treatment of Black soldiers post-war. Note the date; less than 10 days after its publication would be the Jacksonville Mutiny which involved the soldiers of the 3rd USCT, Sgt. Harmon's regiment (and Sgt-Major James' regiment) which had been recalled from Gainesville to Jacksonville to be mustered out:

> *Since the surrender of the troops in Florida by General Samuel Jones, and during the actual existence of the Rebellion, we have been told by our commanding officers on the eve of battle to forget old grudges and prejudices, and fight like men for a common cause, meaning for us not to let the unjust and cruel treatment of the officers to the men, influence us to a disregard for our duty to our common country. But now there is nothing of the kind of fear, the officers having the feeling that they have nothing now to fear from stray bullets, are exercising all the arrogance and despotism that their power gives them, and what appeals has an enlisted man if he applies for redress to the superior officer? It can only be endorsed through the officer who is his worst enemy, whose endorsement will be, as a matter of course, the most detrimental to the interest of the soldier. Now we have the tying up of the thumbs of which Mr. Green* [the writer of the letter to which Sgt. Harmon is responding] *speaks, on the public streets of the town, and what is called riding the horse, which is two upright posts set in the ground, full seven feet high, and three-cornered cross beam, on which men are compelled to sit astride, and other punishment, which even these people, both white and black, are horrified at witnessing, used to slavery and its horrors as they all are. And for what?*

Because some of those stauch [sic] *union men, many of whom wear the uniform of the so-called confederacy, and have not to this day taken the oath of allegiance - but their word is sufficient to condemn any amount of colored soldiers or citizens, for even citizens feel the effect of that most prevalent and baleful disease, negrophobia.*

Postwar Henry S. Harmon, the letter-writing Sergeant of Co. B, 3rd USCT

A reviewer of the Second Edition of this book called attention to the stunning post-war careers of letter writer Sgt. Henry S. Harmon.

Harmon has his own Wikipedia page which depends heavily on a detailed article by Darius J. Young which appeared in the *Florida Historical Quarterly*, Fall 2006, Vol. 85 (2), pp. 177-196.

Harmon was born in Philadelphia, the son of escaped slaves from Virginia. He apparently received a solid education in his youth.

He enlisted on June 30, 1863 with the 3rd US Colored Troops, age 25, standing 5 feet 5 1/2 inches. His pre-enlistment occupation is listed as "sailor" as shown below.

[Company Descriptive Book record for Harry Harmon, Co. I, 3rd Reg't U.S. Col'd Inf., enlisted June 30, 1863 at Philadelphia.]

He was advanced to Corporal on August 2, 1863, and promoted to Sergeant on November 21, 1863, a rapid rise in the ranks.

We have no direct knowledge of his engagement in any particular action during the war, knowing from a letter in *The Christian Recorder* in December 1863 that he took part in the siege of Fort Wagner and Fort Gregg on Morris Island, South Carolina in the fall of 1863, and can assume his participation in operations involving the 3rd USCT in occupying Baldwin, Florida upon arrival in Jacksonville in February 1864, and occupying Palatka in the summer of 1864. When he writes his letter to *The Christian Recorder* about the raid in early April 1865, he is assigned to a likely boring role with an artillery unit in Jacksonville.

In an undated letter to *The Christian Recorder* published on October 21, 1865 - just days before the "Jacksonville Mutiny" noted earlier - Harmon writes about being posted with his unit to Gainesville, Florida beginning June 8, 1865.

Harmon will detail the presence of "negrophobia" pervasive in the officer ranks and in post-war civil affairs, claiming that no one from the 3rd USCT will ever serve in the Army until Black officers are included in the command.

Without knowing precisely, we can assume that Harmon and his unit from the 3rd USCT left Gainesville to return to Jacksonville in late October 1865 to be mustered out, likely being there for the "Jacksonville Mutiny."

He mustered out of the Army with the rest of the 3rd USCT on October 31, 1865. As we see from his final pay receipt, it is noted that he is "to remain in the South."

Indeed, he does, joining with his fellow 3rd USCT colleague (Corporal - Company F) Josiah T. Walls in returning to Gainesville, now as civilians. They will be quite a pair for the next 20 years in Gainesville and Florida politics and beyond. (Josiah Walls has his own Wikipedia page - https://en.wikipedia.org/wiki/Josiah_T._Wall)

In 1868, both Harmon and Walls were elected to the Florida House of Representatives, serving until 1870.

In 1869, Harmon became the first Black person admitted to the Florida Bar. Not attending any law school, Harmon diligently studied the law by himself. An option had been created for Black petitioners to circumvent the standard bar examination and simply be designated as "qualified" and admitted to the bar. Declining the easier route, Harmon stood for the standard open bar examination to gain his admittance. Harmon would form a prosperous law practice with Josiah Walls and Marylander William U. Sanders.

Harmon and Walls would continue to be noteworthy figures in Alachua County and in Florida with Walls becoming among the first Black men from Florida to be elected to the US House of Representatives.

The smart and talented Harmon would hold a variety of civil service positions, including becoming the first and only Black person to be Chief Clerk of the Florida House of Representatives, as well as holding several business occupations.

The reader is invited to read a fuller account of Harmon's post-war activities on his Wikipedia page – https://en.wikipedia.org/wiki/Henry_Harmonand in the in-depth article by Darius Young published in *Florida Historical Quarterly* noted at the beginning (go to https://www.jstor.org/stable/30150703 and register to gain free reading access to the article from JSTOR).

Local historical accounts

As mentioned in the Introduction, there is the so-called "Ocklawaha River Raid Re-enactment." It is a fun event that bears no connection whatsoever to anything real and historical. It only becomes a matter of historical concern when one of their leaders speaks to the media and tries to suggest the event has some historical basis, causing the individual to say all kinds of nonsense. If their spokesperson would simply admit that it is a fun, generic Civil War battle re-enactment with no historical connection, then historical confusion would not be sown.

However, there are local historians whose work deserves some scrutiny.

Local historians in Marion County, Florida, namely the Marion County Historical Commission, a public entity under the Marion County Commission, whitewashed the raid account rendered on their historical marker into another triumph for Capt. Dickison as the good Confederacy apologists they have been. (They have plenty of company in spinning the Confederate "Lost Cause" apologist's revision of history.)

The marker's text is a curious piece of work that deserves closer examination.

Its jumbled and fumbled nature may have been the innocent by-product of history being written by a committee with each member adding their favorites facts, lore, and spin. The final text is inchoate and quite embarrassing in its faultiness.

On the other hand, it gives evidence of having been a purposefully designed narrative with the intent to illustrate a racist trope in the guise of an objectively historical account.

You can judge for yourself.

The first half of the text is dedicated to the Marshall family:

A short distance north of here stood the sugar plantation of Jehu Foster Marshall, established in 1855. At the start of the Civil War in 1861, Marshall was named a colonel in the Confederate Army and soon commanded one of General Wade Hampton's infantry units, the 1st South Carolina Rifles. Colonel Marshall was killed during the Second Battle of Manassas in August 1862. The plantation continued in operation under the supervision of his widow, Elizabeth Anne DeBrull Marshall, until March 10, 1865, when Union troops staged a surprise raid. The Marshall Plantation and the sugar mill were burned to the ground.

One might think several things after reading this first half of the marker text.

First, one might believe that the plantation was itself something historically important. While it was substantial, it was not exceptional; there were many plantations in Marion County of similar or larger scope and they have all vanished and been largely forgotten. Indeed, the only thing historically significant about the plantation is that it is the only one burned to the ground in the Union raid in Marion County.

Second, with the interest in the plantation's owners, one might believe that they were essential figures in Marion County. Rather, the Marshalls were absentee owners of this sugar plantation in southeast Marion County as well as a cotton plantation in northwest Marion County. The Marshalls lived in South Carolina as Col. Marshall's military unit suggests. Perhaps they visited once a year or so to check on things, but they had virtually no involvement in Marion County.

If the plantation was nothing special, and the Marshalls had little to do with it, why all of this information about them? To the Confederacy apologists composing the text, the plantation and its owners were important enough to take up half the text of the marker.

However, if we were to envision this as a picture painted by Confederacy apologists, we would have a grieving widow, victimized by the war in the loss of her brave husband, who now has her plantation destroyed by Union soldiers as well.

The next sentence is particularly troubling on several levels. Here is the text:

> *The raid was conducted by elements of the 3rd United States Colored Infantry, led by the black Sergeant Major Henry James.*

First, this specific identification of a unit participating in the raid and the specific naming of its leader can *only* be gleaned from Union records, such as those published in the massive 128-volume *War of the Rebellion: Official Records of the Union and Confederate Armies* published in 1897 (and available online – see the Brief Bibliography for the full work), which provides this information about Sgt-Major James and the 3rd USCT. Yet any other information about the raid from these Union records was deliberately excluded from the marker's narrative.

Realizing that this information could only have come from research that probed Union records, a sign of some considerable research effort, I was disabused of the notion that the text writers in the late 1990s simply did not have access to the records now available online to a researcher like me. One cannot believe that the text writers did not have access to records when they were able to present information like this. They selectively took only the information that they wanted and disregarded the rest.

Second, why did the local historians include only this information? It has only one purpose: to identify the raiders as Black soldiers, emphasized by specifically identifying Sgt-Major James as "black," as if being identified already as "elements of the 3rd United States Colored Infantry" was insufficient. Ironically, it is this information that led journalist Rick Allen to dig deeper into the historical records.

Again, taking the Confederacy apologist's perspective, the picture of the grieving widow having her plantation destroyed is supplemented by revealing the culprits of this arson (and thievery) not only as Union invaders but as Black men.

The rest of the marker text is a rather embarrassing mess from the perspective of historical methodology which, as said above, could have been avoided had the text writers seriously consulted with any of the available works from the Confederate side. It doesn't seem they were serious about historical research and were perhaps more interested in telling a story with a preferred narrative as opposed to the readily available historical record. Here is the rest of the text:

> *The Ocala Home Guard pursued the Union force and during the running battle, two of the home guard members were killed. After crossing the Ocklawaha River, the raiders set fire to the bridge. Company H, 2nd Florida Cavalry, led by Captain J.J. Dickison, encamped at nearby Silver Springs, soon gave chase and succeeded in driving the Union troops into St. Augustine, and reclaiming all property seized during the raid.*

First, there was no "running battle." There was a brief firefight in the scrub about 12 miles from the bridge. As Sergeant Harmon related in his letter, the only ones "running" were the Home Guardsmen fleeing into the scrub. There was no earlier or further exchange of gunfire between the Home Guard and the raiders, and Dickison would never catch up to the raiders. No "running battle" occurred, but it sounds dramatic.

The mention of a "running battle" in the context suggests that the Home Guard met the raiders at the plantation and that the "running battle" occurred from the plantation to the bridge. It wrongly places the action that occurs much later in the scrub as happening between the plantation and the bridge.

The distance between the plantation and the bridge was only about one mile. The Home Guard could never have received the raid report from the plantation at its location in Ocala, assembled its force, and gotten to the plantation in the half-hour or so it took for the raiders to complete their mission at the Marshall Plantation and cross the Ocklawaha River.

Second, Captain Dickison was not "encamped at nearby Silver Springs." Dickison had never encamped at Silver Springs and would never have had any reason to do so; it was far, far away from any action. In 1865, he was encamped in the area of Waldo. It suggests that Dickison was nearby and directly encountered the Union raiders when, in fact, he never encountered them.

Third, "driving the Union forces into St. Augustine" misses the points that a.) the St. Augustine garrison was the raiders' goal, and b.) Dickison never caught up to them.

Fourth, "reclaiming all property seized during the raid" is false as the Union record reveals; a wagon, horses, mules, and over 70 liberated slaves made it to safety and new life in St. Augustine.

Further, the phrasing is grotesquely, tragically, embarrassingly racist.

Bear in mind that these "historians" in 1999 when the marker was installed had the despicable temerity to state "all property," knowing fully that this included not only wagons and horses but also slaves. Yes, there is no mention of slaves *anywhere* in the marker text, except at the end where the slaves are implicitly lumped together with wagons and horses as likewise "property."

Let us finish the contrived picture that may have been painted by these Confederacy apologists. We had a grieving war widow who was victimized by having her plantation burned to the ground by thieving Black arsonist invaders, a deeply sad and troubling picture from the perspective of Confederacy apologists.

To turn this dour narrative into something inspiring, the text writers portray the Home Guard rushing to the plantation where the militia confronts the Black thieves and arsonists, suffering two fatalities for their heroic effort. As the Black invaders flee, they burn the bridge. The pursuit is then joined by Capt. Dickison, the local hero from Marion County, who saves the day by recovering "all property" from the thieving Black raiders, sending them scurrying empty-handed back to St. Augustine.

The reader of this marker text may have come to the same conclusions that I had in my first encounter with it. I thought that the widow Marshall lived at the sugar plantation she owned in Marion County. Black Union soldiers burned it down, learning later that they apparently stole some items. The raiders were beset by the Home Guard or militia. The raiders escaped across the bridge and then burned the river bridge, halting the Home Guard and the "running battle." The chase was then taken up by Capt. Dickison encamped nearby. He pursued them all the way to St. Augustine and recovered everything that was taken.

Considering the marker's depiction, I regarded the raid as rather pointless with nothing really being gained by it with unnecessary destruction of the widow's property so late in the war. How many others have arrived at a similar conclusion? It seems to reflect the message that the marker's text writers intended.

Of course, it is a remarkable string of lies. The text writers have foisted a fictional account of vindication-via-hero-Dickison, portraying the raid as a vicious and pointless exercise by Black men in uniforms running amok, stealing and burning, compounding the hardship of the grieving war widow in a raid that failed and produced nothing.

This fake narrative was preferable to admitting that, historically speaking, the fox - Dickison - was outfoxed. By Black soldiers no less. The marker's narrative is simply an homage to the Confederacy using Lost Cause revisionism.

The work of local "Lost Cause" historians to portray the Confederacy and its despicable White supremacist ideology as a noble and honorable cause continues to the present day. There are no qualms about manipulating accounts to polish the violence, oppression, depravity, and corruption of the slave-based economy and the racist slavery state while also seeking to diminish, deny, and corrupt the achievements of Black people.

The hubris of these Confederacy apologists ironically left a trail of breadcrumbs leading to the Union raiders. Journalist Rick Allen picked up the breadcrumb trail and his research revealed a more accurate account of the raid in the discovery of the letter by Sergeant Harmon.

Still, decades of sharing deliberate misinformation have corrupted this important account of Black history for the community. It will take many years to replace the long-accepted fraudulent account with the more accurate version provided here.

Hopefully, readers of this book will have a better understanding of the long and difficult journey of the Black soldier and of this fascinating, obscured, and misrepresented raid, celebrating the makers of Black history and their daring raid on the Marshall Plantation.

Brief Bibliography

A variety of resources were consulted which provided an abundance of information. These are some references that an interested reader could pursue in further researching this topic.

Primary Sources:

Dickison, Col. J.J. – *Confederate Military History, Vol. XI – Florida* – ed. Clement Anselm Evans, Confederate Publishing Company: 1899

Link: https://www.perseus.tufts.edu/hopper/text?doc=Perseus%3Atext%3A2001.05.0256%3Achapter%3D6%3Apage%3D133

also recorded in:

Dickison, Mary Elizabeth – *Dickison and His Men: Reminiscences of the War in Florida*, Courier-Journal Job Printing Company: 1890, pp. 208-209

Link: https://books.google.com/books?id=8BcTAAAAYAAJ&pg=PA128&source=gbs_toc_r&cad=4#v=onepage&q=Marshall&f=false

Harmon, Sgt. Henry S. – *The Christian Recorder*, letter dated April 3, 1865 (pub. April 22, 1865)

Link: https://archive.org/details/christianrecorder_1865_v5_no13_to_25/mode/1up - - in the viewer at this link, go to page 14 of 48

Other Sources:

Allen, Rick – *The Mystery of the Marshall Plantation: Fallen Sign, Complicated Past* in the Ocala Star Banner, Feb. 12, 2016

Link: https://www.ocala.com/story/news/local/2016/02/12/mystery-of-marshall-plantation-fallen-sign-complicated-past/31968157007/

Cook, David, *Historic Ocala: The Story of Ocala & Marion County*, HPN Books: 2007

Link: https://books.google.com/books/about/Historic_Ocala.html?id=ppk8qVkQwP8C

Dobak, William S. - *Freedom by the Sword: The US Colored Troops, 1862-1867* – Army Historical Series, Center for Military History, Washington, DC – esp. p. 87 (p. 107 in PDF)

Link to download PDF: https://history.army.mil/html/books/030/30-24/index.html

Emilio, Luis F. (Captain) – *Brave Black Regiment: History of the Fifty-fourth regiment of Massachusetts volunteer infantry 1863-1865* – Boston Book Company: 1891

Kindle version $.99 – https://www.amazon.com/Brave-Black-Regiment-Fifty-Fourth-Massachusetts-ebook/dp/B071FN1MY4/

Glathaar, Joseph T. – *Forged in Battle: The Civil War Alliance of Black Soldiers and White Officer,* The Free Press-MacMillan: 1990

Johnson, William B. D., Clerk, Company A, 3rd US Colored Troops – *The Christian Recorder*, letter dated April 7, 1865 (pub. April 29, 1865)

Link: https://archive.org/details/christianrecorder_1865_v5_no13_to_25/mode/1up -- in the viewer at this link, go to page 18 of 48

Koblas, John J. – *J.J. Dickison: Swamp Fox of the Confederacy*, North Star Press of St. Cloud, Inc.: 2000

Koblas, John J. – *The Swamp Fox and the Columbine*, North Star Press of St. Cloud, Inc.: 2003

Ott, Eloise Robinson, and Louis Hickman Chazal – *Ocali Country, Kingdom of the Sun: A History of Marion County Florida,* Marion Publishers: 1966, esp. p. 83

Schafer, Daniel L. – *Thunder on the River: The Civil War in Northeast Florida*, University Press of Florida: 2010

War of the Rebellion: A Compilation of the Official Records of the Union and Confederate Armies, Government Printing Office, Washington, DC: 1897 – most of the 128 volumes are available in a searchable database online at: https://ehistory.osu.edu/books/official-records

Catalog of Images

These images are listed according to the order of their appearance in the book.

Rick Allen – used with his permission

Masthead of The Christian Recorder – clipped from Internet Archive at archive.org

Secretary of War Simon Cameron - public domain

General Benjamin Butler – public domain

Map of border states – Wikipedia – unattributed – "USA Map 1864 including Civil War Divisions"

Lincoln meets McClellan – public domain - by Alexander Gardner

"Fatigue duty for Black soldiers after the Cold Harbor battle (June, 1864) – public domain

3rd USCT battle flag – Wikipedia

Benjamin Chew Tilghman (post-war) – public domain – by F. Gutekunst

Captain William Mathews – Kansas Historical Society at National Park Service – nps.gov

General David Hunter – public domain

James Montgomery in 1858 – public domain

Harriet Tubman – woodcut – public domain

Combahee River raid – drawing – public domain

Charleston Harbor fortifications – public domain – Harper's Weekly with labels added by the author

Fort Wagner interior and exterior views – public domain – at National Park Service – nps.gov

Fort Wagner Battle Map – American Battlefield Trust – Map prepared by Steven Stanley

Colonel Robert Gould Shaw – public domain

General Quincy A. Gillmore – public domain

Images of Jacksonville in 1865 – three images – public domain

General Truman Seymour - public domain

John Hay in 1862 – public domain

General Pierre T. Beauregard – public domain

General Joseph Finegan – public domain

Confederate torpedo for rivers – public domain

Captain J. J. Dickison – public domain

Spencer 7 diagram – public domain

Attack on the USS Columbine – drawing – public domain

Map – 1865 Florida Map – public domain – with mark-up by the author – ***Dickison's movements in February 1865*** - *Waldo to Braddock Farm to Waldo to Station Four, Levy County*

Flatboat carrying a wagon on the St. Johns River – public domain

Civil War-era pontoon boat – public domain

Map – 1865 Florida Map – public domain – with mark-up by the author – **Raider movements from March 7 to March 10**

Postcard: wild banks of the St. Johns River - public domain

1928 Sharpes Ferry Bridge, replaced in 2012 – by C Hanchey-0 2-2011-bridgehunter.com-BH Photo # 194176

Photo of a sugar plantation operation – postcard – public domain

Sugar cane grinding in 1928 - public domain

Map – 1865 Florida Map – public domain – with mark-up by the author – **Movements to midnight March 10-11**

Map – 1865 Florida Map – public domain – with mark-up by the author – **Movements to 6 am on March 11**

Map – 1865 Florida Map – public domain – with mark-up by the author – **Movements to 12 noon on March 11**

A road through the scrub - public domain

Map – 1865 St. Johns County, Florida Map – public domain – with mark-up by the author – **Movements to after midnight on March 12**

Union picket camp at Bull Run in Virginia - public domain

Old gates of St. Augustine from 1865 - public domain

General Gillmore's Hilton Head SC HQ – public domain

Major Martin Robison Delaney - public domain

C. Powell; B. Obama; L. Austin – official pictures compilation

Sergeant-Major Henry James headstone – unknown source

John C. Breckinridge – public domain

Hanging a prisoner by the thumbs – public domain

Appendices

Here are the full texts of the significant items, publications, and correspondence mentioned in the book.

No. 1 – Marshall Plantation Historical Marker text *by Marion County Historical Commission, 1999*

A short distance north of here stood the sugar plantation of Jehu Foster Marshall, established in 1855. At the start of the Civil War in 1861, Marshall was named a colonel in the Confederate Army and soon commanded one of General Wade Hampton's infantry units, the 1st South Carolina Rifles. Colonel Marshall was killed during the Second Battle of Manassas in August 1862. The plantation continued in operation under the supervision of his widow, Elizabeth Anne DeBrull Marshall, until March 10, 1865, when Union troops staged a surprise raid. The Marshall Plantation and the sugar mill were burned to the ground. The raid was conducted by elements of the 3rd United States Colored Infantry, led by the black Sergeant Major Henry James. The Ocala Home Guard pursued the Union force and during the running battle, two of the home guard members were killed. After crossing the Ocklawaha River, the raiders set fire to the bridge. Company H, 2nd Florida Cavalry, led by Captain J.J. Dickison, encamped at nearby Silver Springs, soon gave chase and succeeded in driving the Union troops into St. Augustine, and reclaiming all property seized during the raid.

No. 2 – Sergeant Henry S. Harmon's letter – *text lightly edited* – in The Christian Recorder, April 22, 1865

THE JACKSONVILLE EXPEDITION.

MR. EDITOR: - The many expressions of heartfelt sympathy and kindness that are borne to us on every northern breeze, from our friends at home, for the colored soldier, compels us to let nothing transpire that we think will be interesting to them to know, without giving them due information of it, so far as lies in our power.

In view of this fact I take the liberty of your columns, to present for their perusal an account of an expedition, which left Jacksonville under the command of Sergeant-Major Henry James, 3d U.S.C.T., on the night of the 7th of March consisting of sixteen (16) of the 3d U.S.C.T., six (6) men of the 34th U.S.C.T., and seven colored citizens, and one (1) of the 107th O.V.I. Thirty (30) men in all.

After waiting some time for darkness to throw her pall over the scene, the commander gave the order to push off. The party then moved up the St. John's River, in pontoon boats to Orange Mills, where he landed with ten men and skirmished the country to a point near Pilatkia, where the boats met them, and seeing all well, he again skirmished to what is called Horse Shoe Landing, said to be 100 miles from Jacksonville, which brought them well up in the day: having fatigued themselves considerably, they remained in the swamp until the boats came up, about nine o'clock in the evening, when he embarked again and proceeded to what is called Fort Gates.

He then ordered the boats pulled close into the shore under cover of the dense swamps, and proceeded with the whole force across the country to the Oclawaha River, to what is known as Marshall's plantation. Here was one of the objects of the expedition reached without serious opposition, and almost in the heart of the enemy's country, and as yet quite unknown to him.

Here the expedition captured some 25 horses and mules, burnt a sugar mill, with 85 barrels of sugar, about 300 barrels of syrup, a whiskey distillery, with a large amount of whiskey and rice, and started on their return, bringing along 95 colored persons, men, women and children, recrossed the Oclawaha River, burning the bridge.

Six men then were detached from the command and sent under charge of Sergeant Joel Benn, of Co. B, 3d U.S.C.T., with Israel Hall, scout, to Hawley plantation, where they were attacked by a small body of rebels, and Sergt. Benn was killed, shot through the heart, Henry Brown, scout, wounded, and Israel Hall, chief scout, captured, as was another citizen named Ben. Gant, the others being compelled to return to the main body.

Their troubles had now commenced in earnest, this being the second fight of the day, for having to charge the bridge in going to Marshall's and killing three rebels had only stirred them up, but they pushed on, for much of their success depended on their speed.

But when within about twenty miles of the St. John's River, the enemy numbering about fifty men well-mounted, came down on them, calling on them to surrender, or suffer themselves to be hanged.

But there was another alternative which he, the enemy, did not think of, and which the Sergeant Major, who, by the way is not a surrendering man, resolved to take, which was to fight them awhile first.

Seeing this, the enemy prepared himself to make it warm for the little band of colored men. Breaking to the right and left under cover of a hill, they dismounted and formed their line of attack, and came over the crest of the hill, in quite an imposing array to find the little band of seventeen men, (the balance being left to guard some prisoners and the avenues of retreat,) deployed, as skirmishers to meet them, covered as much as possible by the trees. But on they came.

And every man selecting his man, when they were near enough for every man to make sure and waste no ammunition, Sergt. James gave the command to commence firing, and for awhile nothing was heard but the sharp crack of the soldiers' rifle and the louder roar of the citizens' fowling-piece, blended with the yells of their wounded and dying. The firing on the part of our men was good, as was shortly proved, for the enemy suddenly broke for their horses, when our men, leaving their cover, dashed in among them with the bayonet and clubbed guns, scattering them in every direction, leaving some 20 of their men dead and a few wounded.

Finding the way clear again, Sergt. James, on summing up, found the woods had afforded them such good covering that he had only two men wounded, and after taking possession of the best of their horses, (although the enemy suffered so severely, he showed himself to be no mean marksman, as numerous holes in our men's clothing amply testifies, among which, a hole through the commander's cap, caused him to withdraw his head from a dangerous position,) he again took up the line of march for the St. John's, having to abandon one wagon on the way, and soon reached the river and commenced crossing at 12 o'clock on the night of the 10th, and at daylight on the 11th had all across except 9 horses, when the enemy coming up made [it] impossible to recross, consequently had to leave them.

They then destroyed the three boats which they had used, and pushed on towards St. Augustine, and by the time they had got one day's start, Dickerson's guerrilla cavalry were in full pursuit, and, when within seven miles of St. Augustine, the enemy overtook some of the colored people, who were unable to keep up, 19 in number. The remainder of the party reached St. Augustine on the 12th inst., in safety with the wounded, 4 prisoners, 74 liberated slaves, 1 wagon, 5 horses, and 9 mules, having travelled over 200 miles of the enemy's country, doing without food for 3 days, and 100 miles of our own country, in five days and nights, reaching Jacksonville last evening, the 19th inst., with all their booty.

This expedition reflects great credit on Sergt. Major James, for the masterly manner in which it was commanded, and gives further proof, that a colored man with proper training can command among his fellows and succeed where others have failed. And a great deal is due to the men for their good behavior, and steadiness, and obedience, and if it were not for occupying too much of your space, which

I fear I have done already, I would give their names, but that at some other time.

I am still an ardent lover of my race, and a soldier.

H.S.H.,

Sergeant Co. B., 3d U.S.C.T.

Battery McCrea, Jacksonville, Fla.,

April 3d, 1865.

No. 3 – Captain J. J. Dickison's account in Confederate Military History, *Vol. 11, published in 1899, p. 133-134*

On March 15th Captain Dickison reported subsequent operations in his field as follows:

On the evening of the 10th inst., I received information from Marion county, through Col. Samuel Owens, that the enemy was advancing by way of Marshall's bridge and had advanced 12 miles in the interior, burning the bridge. I immediately ordered out my command and in two hours was in rapid march in that direction. While near Silver Springs a courier reached me with a dispatch, stating that the enemy had burned the Ocklawaha bridge and were retreating toward the St. John's river. I then ordered my command to march back in the direction of Palatka, and sent an advance guard to have the flatboat in readiness for us to cross the river. On arriving at the river the wind blew very strong, which delayed our crossing about ten hours. After much difficulty, hard labor and great peril, we succeeded in crossing

50 of my command, leaving the remainder with one piece of artillery to guard and picket other points on the river. Hearing, on my arrival at Palatka, that the enemy had gone up the river in barges, I marched all night and at times at half speed and reached Fort Peaton, 7 miles from St. Augustine, where I overtook four negroes. We continued at fast speed toward the city and within a mile of their picket line, and captured twenty more, also a wagon and six ponies. Three of these ponies have since been claimed by citizens and delivered to them. The enemy, on hearing we were in pursuit of them, left wagons, mules and provisions at the river, where they had crossed near Fort Gates.

The march was truly a hard one. We marched four days and nights with but little forage or provisions. My men were resolved, and showed a determination to pursue the enemy to the very gates of the city. The negroes, twenty-four in number, with the wagons and mules captured, belonged to Mrs. Marshall, of Marion county. The raiding party on reaching her plantation destroyed 200 hogsheads of sugar. Some of our militia met them, and in an engagement two of our men were killed. Had information reached me earlier they would have been overtaken with their rich spoils before reaching the river. All praise is due these noble, gallant men for their unflinching spirit and resignation, having endured every hardship without a murmur.

No. 4 – General Orders from the Office of Major General Gillmore, *by Assistant Adjutant General W. L. M. Burger* (General Gillmore was occupied with the flag-raising ceremony in Charleston at this time.) published in War of the Rebellion: Official Records of the Civil War

https://ehistory.osu.edu/books/official-records/100/0190

GENERAL ORDERS,

HDQRS. DEPARTMENT OF THE SOUTH, Numbers 42. Hilton Head, S. C., April 12, 1865.

On March 7, 1865, a party of colored soldiers and scouts, thirty in number, commanded by Sergt. Major Henry James, Third U. S. Colored Troops, left Jacksonville, Fla., and penetrated into the interior through Marion County. They rescued 91 negroes from slavery, captured 4 white prisoners, 2 wagons, and 24 horses and mules; destroyed a sugarmill and a distillery, which were used by the rebel Government, together with their stocks of sugar and liquor, and burned the bridge over the Oclawaha River. When returning they were attacked by a band of over fifty cavalry, whom they defeated and drove off with a loss of more than thirty to the rebels. After a long and rapid march they arrived at Saint Augustine on March 12, having lost but 2 killed and 4 wounded. This expedition, planned and executed by colored men under the command of a colored non-commissioned officer, reflects great credit upon the brave participants and their leader. The major-general commanding thanks these courageous soldiers and scouts, and holds up their conduct to their comrades in arms as and example worthy of emulation.

By command of Major General Q. A. Gillmore:

W. L. M. BURGER,

Assistant Adjutant-General.

No. 5 – Letter of Mrs. H. B. Greely, dated March 18, 1865 *as provided in "We Are Truly Doing Missionary Work": Letters from American Missionary Association Teachers in Florida, 1864-1874 by Joe M. Richardson in* The Florida Historical Quarterly: *October 1975, published by The Florida Historical Society, pp. 178-195, esp. p. 186-187*

St. Augustine, Fla.

Mar. 18th-64

Dear Br. Whipple

... Twenty five Colored men from Col. Tighlman's *[sic – Col. Benjamin Chew Tilghman, commander of the 3rd USCT until he was relieved of command in February 1865]* regiment stationed in Jacksonville went out on raid last week in this state, and brought into this City, Seventy contrabands. They were the most destitute objects I ever saw. Many of them almost entirely naked. The teachers of the F. R. Ass. [Freedmen's Relief Association] having some money resulting from the sale of books & c. and other means which we could raise, have been very busy this week in making up clothes for the women and children, and we shall soon have them in comfortable condition to remain here, or go elsewhere as Government may see fit to dispose of them. The raiders also brought in some fine horses and mules.

They would have brought more people and more booty had they not been betrayed by a girl on the plantation where they had killed the Overseer, & burned the sugar mills with a quantity of sugar syrup & whiskey and the body of the Overseer in the sugar house.

This betrayal brought upon them a portion of Dickenson's [sic] Guerilla army about seventy, with which they had a fight on Friday P.M. before they reached here on the Sab. Following. They killed the Capt. And 27 of his men, wounding eleven and capturing four whom they brought in with them, making forty three, out of seventy of the rebels, and lost of their own number on the guide who was captured. Doesn't this show Negro valor?

And they claim a little humanity, as they say they left several of the rebels so severely wounded and alone, as their companions had fled, they thought duty to go back, a few of them, and finish them. They say when the parties met they charged upon the rebels in the name of "Fort Pillow."

Dickenson, the "John Morgan" of this Guerilla band, is highly enraged and determined to have these Col. Men if possible. So the Tues. night following the Sab. After they came in, at two o'clk we were awakened by the report of a heavy gun at the Fort and a cry from the guard – "two o'clock and alarm in the Camp." It was found that several of the rebels had crossed the river about a half mile in the rear of the City and others were on the way in their "dug-outs," but they were scattered leaving their boats behind. We are about being reinforced and shall not probably fall into their hands ....

Very Truly Yours, [Mrs.] H. B. Greely

No. 6 – Newspaper article about the raid - Jacksonville Union, March 18, 1865, p. 1

THE WAR IN FLORIDA

A Raid into the Interior and its Results

A party consisting of a detachment of the 3rd and 35th U.S.C.T., left Jacksonville on Thursday the 7th inst., on an expedition into Rebeldom. They landed on the morning of the 8th at Orange Mills. The party here divided, a portion going by boats and a portion by land to oposite Pilatka [sic]. Here they again united and proceeded up the river to Fort Gates where they landed and struck out into the interior.

They soon reached the plantation of Mr. Mason. Here they found two rebel soldiers planting whom they made prisoners. They also secured five stand of arms. They next visited the plantation of Mr. Marsh. Here they captured two more soldiers, one horse, one mule, and six stand of arms. At the plantation of Mr. Williams they captured two contrabands and one horse. They then struck the bridge over the Ocklawaha river where they encountered two pickets belonging to Capt. House's company, who fled to Col. Marshes' plantation where they were overtaken and shot.

At this place they made a haul of 21 horses and mules with their equipments, and 75 contrabands, and burned 75 hogsheads of sugar, 350 barrels of syrup, 400 barrels of whiskey, and the still and sugarworks. This was accomplished by three men. The rest of the party had been left to guard the bridge. The expedition then set out on its return. At Lake Church hill they were overtaken by Captain House's cavalry, numbering some 32. A fight took place which lasted about two hours.

The rebels were defeated with 26 killed and two wounded. The loss on our side was one killed and two wounded. The casualty on our side was occasioned by the treachery of a rebel who had surrendered and afterward fired upon his captors, killing one and wounding another. The party succeeded in crossing the St. Johns river and reaching St. Augustine on Sunday the 12th, with all their booty except five and two prisoners, who were lost in the fight; having marched over 300 miles. No official report has been received at headquarters as yet, of the affair. When the official report is received more particulars will probably be learned.

(Note: There was no further mention of the raid in subsequent issues of the Jacksonville Union.)

No. 7 – Letter to the Editor of the Quincy (FL) Dispatch, March 22, 1865 as reprinted in Dickison and His Men by Mary Dickison, pp. 209-210

FROM EAST FLORIDA

[Correspondence of the "Dispatch," Quincy, Fla.]

Baldwin, East Florida, March 22, 1865.

Editor Dispatch:

Captain Dickison recrossed the river St. John's a few days since with twenty-four negroes, several deserters, wagon, mules, etc., which he had recaptured from the enemy within a mile of St. Augustine. The negroes and wagons belonged to Mrs. Marshall, of Marion county.

She is the widow of Colonel Foster Marshall, who commanded one of the South Carolina regiments of cavalry of Hampton's Legion, and was killed in 1862, in one of the battles around Richmond. She was one of the largest sugar planters in East Florida, and made, last year, at least two hundred hogsheads, all of which was destroyed by the raiders, except twenty, which they endeavored to carry with them, and pressed her mules and wagons for that purpose. A portion of these bold raiders were met by some of our militia, and, in an unfortunate engagement, two of our men were killed.

Captain Dickison, receiving information of their raid, and that they had retreated in the direction of the St. John's river, started in pursuit of them. Pursued by this heroic and intrepid officer, with a detachment of his brave men, they [the raiders] recrossed the river, burned the bridge, and had nearly made their escape, but were overtaken in the very suburbs of St. Augustine.

The cavalry, discovering themselves so closely pursued, put spurs to their horses and galloped into town, leaving their 'colored brethren' to fall a prey to the 'War Eagle' of Florida. He made them right-about, and marched them back to the 'old plantation home,' having it in his power to restore with them much stolen property to the owners.

E. O.

[Captain Dickison has not only the applause and thanks of every true man and woman in Florida, but the still higher satisfaction which attends him in the consciousness of having done his duty faithfully from the beginning of the war to the present hour, and reflected honor upon his country in the noble station assigned him. — Editor.]

No. 8 – Sergeant Henry S. Harmon's Letter published October 21, 1865 in *The Christian Recorder*

Letter from Gainesville, Florida

Mr. Editor:

In your paper of Sept. 9th, 1865, I saw a letter from Fort Bailey, over the signature of William P. Green, and Sir, how glad would I be if I could contradict the statement in it made! But even for my dear friend I cannot. I can only endorse it: honor and justice demand that I and every other colored soldier should or every one from this military division that I have heard speak upon this matter, and in evidence of the fact I will state the experience of our own command, as it has been since we left Jacksonville, for this post *[Ed.-in Gainesville]* on June 8th, 1865. Since the surrender of the troops in Florida by General Samuel Jones, and during the actual existence of the Rebellion, we have been told by our commanding officers on the eve of battle to forget old grudges and prejudices, and fight like men for a common cause, meaning for us not to let the unjust and cruel treatment of the officers to the men, influence us to a disregard for our duty to our common country. But now there is nothing of the kind of fear the officers having the feeling that they have nothing now to fear from stray bullets, are exercising all the arrogance and despotism that their power gives them, and what appeals has an enlisted man if he applies for redress to the superior officer? It can only be endorsed through the officer who is his worst enemy, whose endorsement will be, as a matter of course, the most detrimental to the interest of the soldier. Now we have the tying up of the thumbs of which Mr. Green speaks, on the public streets of the town, and what is called riding the horse, which is two upright posts set in the ground, full seven feet high, and three-cornered cross beam, on which men are compelled to sit astride, and other punishment, which even these people, both white and black, are horrified at witnessing, used to slavery and its horrors as they all are. And for what? Because some of those stauch *[sic]* union men, many of whom wear the uniform of the so-called confederacy, and have not to this day taken the oath of allegiance - but their word is sufficient to condemn any amount of colored soldiers or citizens, for even citizens feel the effect of that most prevalent and baleful disease, negrophobia. Negro citizens although they have been the only true and avowed friends of the United States Government in this section of the country, are still compelled to feel that they are black, and the smooth oily tongue of the white planter is enough to condemn any number of them to tying up for twenty-four hours *[Ed.-presumably by thumbs]*, or two hours up and one down *[Ed.-p resumably astride the beam]*.

Such, my friends, is what we endure or witness, and if the United States Government ever gets five year men, she will not get them from the veterans of the 3d Regiment U.S.C.T., until she is compelled to give us officers of our own choice, who will be officers and gentlemen . Officers who can sympathize with the enlisted man without regard to color; men who will take into consideration a man's former conduct before punishment.

We are rarely allowed to mingle with the people of color around us; in consequence of which, I have not written to you for a considerable time, although there is considerable interesting matter to be found worth relating to your readers, which would throw considerable light upon many things that at present seem dark to the public mind. Hoping that I have not occupied too much space, I am truly a soldier, and I hope a good soldier.

H. S. Harmon

Co. B, 3d U.S.C.T.

www.ingramcontent.com/pod-product-compliance
Lightning Source LLC
Chambersburg PA
CBHW071213090426
42736CB00014B/2808